Ecological Perspectives on Carcinogens and Cancer Control

Progress in Biochemical Pharmacology

Vol. 14

Series Editor
R. PAOLETTI, Milan

S. Karger · Basel · München · Paris · London · New York · Sydney

Selected Papers of the International Conference, Cremona 1976

International Conference on Ecological Perspectives on Carcinogens and Cancer Control

Volume Editors
C.C. STOCK, New York, N.Y.; L. SANTAMARIA, Pavia;
P.L. MARIANI, Cremona, and S. GORINI, Milan

51 figures and 42 tables, 1978

RC 268.6
I 57
1976

S. Karger · Basel · München · Paris · London · New York · Sydney

Progress in Biochemical Pharmacology

Vol. 13: Atherogenesis. Morphology, Metabolism and Function of the Arterial Wall. Eds: SINZINGER, H.; AUERSWALD, W. (Vienna); JELLINEK, H. (Budapest), and FEIGL, W. (Vienna). X+352 p., 143 fig., 29 tab., 1977
ISBN 3-8055-2761-6

Cataloging in Publication
    International Conference on Ecological Perspectives on Carcinogens and Cancer Control, Cremona, 1976.
    Ecological perspectives on carcinogens and cancer control: selected papers of the international conference, Cremona, 1976.
    Volume editors, C.C. Stock et al. – Basel, New York, Karger, c1978.
    (Progress in biochemical pharmacology, v. 14).
    Conference held under the auspices of the European Institute of Ecology and Cancer – Italian Section and other agencies.
    1. Carcinogens – congresses 2. Carcinogens, Environmental – congresses 3. Ecology – congresses 4. Neoplasms – prevention & control – congresses I. Stock, Charles Chester, 1910 –, ed. II. European Institute of Ecology and Cancer. Italian Section III. Title IV. Series.
    W1 PR666H v. 14/QZ 202.3 I61e 1976
    ISBN 3-8055-2684-9

All rights reserved.
No part of this publication may be translated into other languages, reproduced or utilized in any form or by any means, electronic or mechanical, including photocopying, recording, microcopying, or by any information storage and retrieval system, without permission in writing from the publisher.

©   Copyright 1978 by S. Karger AG, 4011 Basel (Switzerland), Arnold-Böcklin-Strasse 25
    Printed in Switzerland by Tanner & Bosshardt AG, Basel
    ISBN 3-8055-2684-9

Contents

Preface VI
STOCK, C.C. (New York, N.Y.): Cancerogenesis, an Ubiquitous Hazard 1
CALVIN, M. (Berkeley, Calif.): Chemical Carcinogenesis 6
LATARJET, R. (Paris): The Impact of Nuclear Technology on the Natural Environment and Human Life 28
FLAMANT, R. (Villejuif): Epidemiological Research on the Relationship between Tobacco, Alcohol and Cancer 36
MALTONI, C. (Bologna): Predictive Carcinogenicity Bioassays in Industrial Oncogenesis 47
GRICIUTE, L. (Lyon): Measurement of Chemical Carcinogens in the Human Environment: the Objectives and Problems Encountered 57
SMITH, K.C. (Stanford, Calif.): Aging, Carcinogenesis and Radiation Biology 70
BOYLAND, E. (London): Significance of Occupational Cancer 76
TÓTH, K.; SUGÁR, J.; SOMFAI-RELLE, S., and BENCE, J. (Budapest): Carcinogenic Bioassay of the Herbicide 2,4,5-Trichlorophenoxyethanol (TC PE) with Different 2,3,7,8 Tetrachlorodibenzo-p-dioxin (Dioxin) Content in Swiss Mice   82
RODIGHIERO, G. (Padova): The Problem of the Carcinogenic Risk by Furocoumarins   94
MUÑOZ, N. (Lyon): Perinatal Viral Infections and the Risk of Certain Cancers 104
BALDWIN, R.W. (Nottingham): Immunology of Rat Hepatic Neoplasia 109
PLESCIA, O.J.; GRINWICH, K.; SHERIDAN, J., and PLESCIA, A.M. (New Brunswick, N.J.): Subversion of the Immune System by Tumors as a Mechanism of their Escape from Immune Rejection 123
LAPIS, K. (Budapest): Morphological and Biological Features of MC-29 Virus-Induced Liver Tumors in Chicken 139
HUEBNER, R.J.; PRICE, P.J.; GILDEN, R.V.; TONI, R.; HILL, R.W., and FISH, D.C. (Bethesda, Md./La Jolla, Calif./Frederick, Md.): Immunoprevention of Leukemia in AKR Mice by Type-Specific Immune Gamma Globulin (IgG)...... 151
MONTESANO, R. (Lyon): The Use of Mutagenicity Tests in Screening Chemical Carcinogens 157
FORNI, G. and CAVALLO, G. (Torino): Requirements for Tumor Antigen Immunogenicity 163

Preface

We are happy to present part of the contributions to the International Conference on Ecological Perspectives on Carcinogens and Cancer Control held in Cremona (Italy) on September 16–19, 1976, under the auspices of the European Institute of Ecology and Cancer – Italian Section, Lega Italiana per la Lotta Contro i Tumori, Fondazione Giovanni Lorenzini, and with the technical contribution of the International Agency for Research on Cancer (IARC).

The philosophy we so much discussed to organize the above-mentioned Conference on such a new but terribly urgent problem in carcinogenicity is expressed in the Introductory Chapter. The picture of our efforts is completed by the contributions published in a supplementum of *Medicine, Biology, Environment, International Bimonthly Magazine*, Belgium.

We hope our volume will prove useful to stimulate future progress in this area of research.

The Editors

Cancerogenesis, an Ubiquitous Hazard

C. CHESTER STOCK

Memorial Sloan-Kettering Cancer Center, New York, N.Y.

I can think of no better way to commemorate the 40th Anniversary of the foundation of the Consorzio Provinciale per la Lotta Contro i Tumori di Cremona than to hold this International Conference on Ecological Perspectives on Carcinogens and Cancer Control. We are greatly indebted to the Italian Section of the European Institute of Ecology and Cancer, to the Lega Italiana per la Lotta Contro i Tumori, and to the Fondazione Giovanni Lorenzini for promotion of this conference. I appreciate very much having the opportunity to participate.

One may well ask how someone who has spent most of his scientific life in cancer chemotherapy has the temerity to participate in a meeting on cancerogenesis. I will ask in return who else has a better reason to advocate the prevention of cancer through elimination or minimizing cancerogens in our environment than one who has spent much of his life trying to find ways of treating cancer with drugs. I know too well the difficulties of finding such agents and how unhappily we see problems associated with those found useful in clinical cancer. Truly, prevention is many times better than cure.

While at Memorial Sloan-Kettering Cancer Center we have not mounted a large program on cancerogenesis since the pioneering work of Dr. WYNDER and his associates on tobacco and carcinogenesis, we do have a number of pertinent investigations as will become more evident in the presentations made at this meeting. My colleagues welcome this opportunity to exchange information in this conference.

Truly, cancerogenesis is an ubiquitous hazard. Experts give us estimates that from 75 to 90% of cancers are caused by factors in the environment. Cancerogens are found in water, air, soil, food and even in medicines. Hardly a day goes by without seeing headlined in our newspapers that something in more or less common use has been found to be or is suspected to be cancerogenic or that some known cancerogenic agent has been found to be a serious contaminant in some new part of our environment.

Typical recent headlines to alarm and shock us on toxic or cancerogenic contamination read like this: 'Home plasterers warned about asbestos in their materials'; 'Kepone spreads in Chesapeake Bay'; 'Mirex being discharged into Niagara River'; 'Poisonous cloud released in Northern Italy'.

Only relatively recently have we awakened to the evidence increasingly recognized of the very large burdens of toxic agents, cancerogens or potential cancerogens that have been released in our environment. Chemicals used to combat our ecological enemies are now making our environment hostile to us.

With increasing recognition of the magnitude as well as the ubiquitous nature of the hazards of cancerogenesis has come the realization of the neglect given to studies in the field of chemical cancerogenesis. While excellent research has been conducted by highly competent investigators, the field has had too little support. Because of the lack of support and partly because of the less glamorous, long-term studies required for cancerogenesis determinations, an inadequate number of investigators have chosen this field so essential to the welfare of all of us.

The concern about this is expressed by a report of the US General Accounting Office which criticizes the Federal Government's efforts in cancerogenesis. I take the liberty of quoting from that report:

'Although up to 90% of human cancer, according to some scientists, is environmentally caused and controllable, federal efforts to protect the public from cancer causing chemicals have not been very effective.

'Many chemicals cause cancer in animals, but federal agencies have trouble determining which also pose a cancer threat for humans because:

'– There are no generally accepted principles concerning environmental causes of cancer.

'– There are no uniform minimum guidelines for testing.

'– Test data are not always complete or appropriate.

'– Scientists cannot accurately predict human response to chemicals on the basis of animal test results.

'NCI is responsible for directing federal efforts and should, with the cooperation of other involved federal agencies, develop a uniform federal policy for identifying and regulating cancer causing chemicals.

'The policy should at least cover:

'– The information needed to regulate cancer causing chemicals

'– Which chemicals should be tested in animals.

'– How tests should be conducted.
'– How results should be evaluated.
'– How human risk can be assessed from animal studies.
'– What factors other than public health should agencies consider.

'Although HEW agrees that a federal policy is needed, it does not agree that a formal effort, headed by NCI, is necessary. GAO believes a federal policy can only be developed with the active support of every involved federal agency, and the NCI director, as head of the National Cancer Program, should coordinate these efforts.

'GAO is also recommending that FDA have all approved and proposed food additives tested for their cancer-causing potential because it had not been requiring data from such tests when the additives were unintentionally added to the food in amounts less than 1 or 2 parts per million. HEW disagrees, saying the risk of cancer is remote and the costs for testing would be substantial.

'Tobacco and tobacco products are on NCI's list of known human carcinogens. For the last two years, the HEW secretary has recommended that Congress give the Executive Branch authority to control hazardous ingredients in cigarettes, such as tar and nicotine.

'GAO suggests that Congress request HEW to prepare a study showing the available options for regulating tobacco products, and the impact each option would consider giving HEW or some other appropriate agency the specific authority to regulate tobacco and tobacco products.'

Parenthetically, these general points apply beyond the US and require international consideration.

It is our hope that this conference will provide at least a few answers to some of the questions raised in that report and for some of the others take at least useful steps along the path to solution of the problems posed. Certainly, the topics covered in this meeting provide encouragement that this will happen.

An example of the increasing recognition of the problem was the 11th Canadian Cancer Research Conference on environmental carcinogenesis held last May, with a meeting of the American Association for Cancer Research and the biennial meeting of the International Union Against Cancer. A number of topics to be discussed in our present conference were considered at the Canadian meeting under the general headings of Factors Influencing Public Policy; Environmental Carcinogenesis for Man; Sensitivity and Monitoring of Human Populations; and Cellular and Molecular Responses to Environmental Carcinogenesis. From reports at that meeting even one not versed in carcinogenesis studies could gain some concepts of the ramifications of the problem. Of the millions of known compounds, many thousands are produced commercially in a variety of formulations. On the toxic substances list of the National Institute for Occupational Safety and Health are nearly 1500 compounds alleged to be cancerogenic in at least one animal species. It has been stated to cost $ 100,000 requiring work on 600–700 animals covering a period of 3 years to test one material for carcinogenicity. Clearly that calls for a reliable method for ordering priorities. Even after demonstrations that a substance is cancerogenic, there may be problems in effecting use of that knowledge as exemplified by the failure to influence adequately smokers to stop their habit. JUKES has cited a problem that comes from a ban on the use of diethyl stilbestrol in the production of cattle. Although some may challenge his figures, they do illustrate the economic problem that banning a carcinogen can create. He estimates the ban would prevent one cancer from DES in 133 years at a cost of $ 9,000,000 yearly and utilization of 7 million tons of corn.

I will now return to a brief review of cancerogenesis studies in our Institute. We are not attempting an in depth coverage of the field but are pursuing individual studies of selected compounds, of a detection procedure, of a special environment, and of an enzymic system.

The need for *in vitro* tests for screening cancerogenic hazards is well known and some have been proposed by various investigators. It may be that the most useful screening will consist of a battery of such tests, if no one test is sufficiently reliable. The value of a sensitive, rapid and readily quantified test was evident in the earlier cited on the magnitude of the problem. Dr. HANS MARQUARDT will report on his cell culture system using mouse fibroblasts to test for carcinogenicity and to study mechanisms of action of carcinogens.

Taking a lead from the clinic, Dr. LUCY ANDERSON has been investigating a special environment, most important I am sure you will agree, through studies of transplacental carcinogenesis. The possible production of tumors in embryos by maternal exposure chronically to either dimethyl nitrosamine or benzo(a)pyrene and in other experiments to oncogenic or potentially oncogenic purines. Extensions of the study will include several factors that might have an influence upon any observed tumor formation.

Studies on chemical carcinogenic agents have been focused in our Institute on three categories of compounds. Dr. GINER-SOROLLA has been particularly interested in the nitrosamines from nucleic acid components that might be formed in our internal environment. He has also been interested in the possibilities of *in vivo* interaction of nitrites in food or saliva with psychotropic drugs, and is conducting studies on nitrosamines with Dr. ANDERSON.

Dr. GEORGE BROWN who, because of illness in the family regrets he cannot be here, has in recent years been very much involved in elucidating information about the oncogenicity of purine-N-oxides. There is the possibility that these can arise in our internal environment, namely *in vivo*. A number of purine-N-oxides and related compounds have been synthesized and tested for oncogenicity. Studies particularly focused on 3-hydroxyxanthine show it is metabolically converted to a sulfate ester which can undergo two parallel competing reactions to yield either a radical or ion. Both of these highly reactive intermediates must be considered as possible proximate oncogens. Studies of the oncogenicity of several specially selected 3-hydroxy purines for correlation with ability to produce either type of intermediate are expected to indicate which may be preferred as the proximate oncogen.

It was HADDOW who first called attention to the dual effects of certain compounds, namely cancerogenic and antitumor activities. Dr. FREDERICK PHILIPS has been conducting studies which emphasize the parallelism of pathologic effects of antitumor agents and chemical carcinogens in cell renewal systems. The further demonstrations of cancerogenicity of additional antitumor agents illustrate that we in our treatment of disease through our medicine may be contaminating the patient's internal environment with carcinogens with the consequent risk of further disease. This is a factor to be considered in cures of at least some forms of cancer especially in younger patients.

Finally, we come to the studies of Dr. ROBERT ANDERSON who has been studying invertebrate animals. This provides an opportunity to determine the impact of environmental carcinogens upon non-mammalian components of the ecosystems. He has observed that mixed function oxidases of two species can metabolize benzo(a)pyrene. He will discuss with you the implications of these studies.

On behalf of all my colleagues, I wish to express our appreciation to the promoters and organizers of this conference for the opportunity to learn from our participation in the conference.

Dr. C.C. STOCK, Vice-President for Academic Affairs, Sloan-Kettering Institute for Cancer Research, Donald S. Walker Laboratory, 145 Boston Post Road, *Rye, NY 10580* (USA)

Chemical Carcinogenesis[1]

MELVIN CALVIN

Laboratory of Chemical Biodynamics, University of California, Berkeley, Calif.

Introduction

Chemical carcinogenesis is a term which is used to describe the fact that many kinds of natural and synthetic chemicals present in our environment could conceivably be components in the triggering, or genesis, of malignancy. It has been known for over 100 years that a component of soot is indeed a principal source of certain kinds of cancer, and it has been known for the last 20 years which component of soot is most active in this regard. Since this early recognition, the number of chemicals (natural and synthetic) which have been designated carcinogens has increased enormously, usually by virtue of some kind of epidemiological study or more recently and, more frequently, by deliberate screening programs with animals of various kinds.

The structure of today's discussion is: First, is there any chemical event, or common chemical property, of these chemical materials which has been recognized and is there any common chemical reaction which they perform; and, secondly, is there any common mechanism of achieving the biological consequences which we know these chemicals have?

Chemical Carcinogenesis

A good deal of progress has been made during the last 2 decades in learning more about a whole variety of organic chemical carcinogens, and a

[1] The work described herein was sponsored, in part, by the National Cancer Institute (through Grant No. 2 POI CA 14828-04), in part by the US Energy Research and Development Administration and, in part, by the Elsa U. Pardee Foundation for Cancer Research.

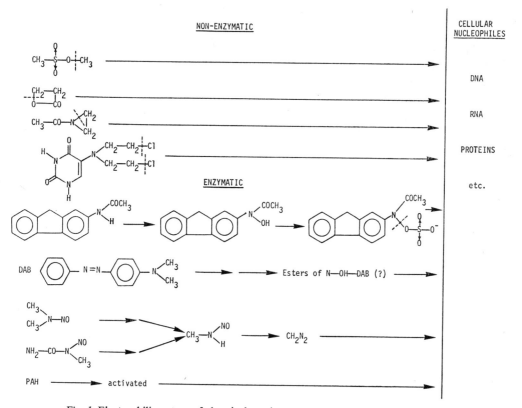

Fig. 1. Electrophilic nature of chemical carcinogens.

rather straightforward view of chemical carcinogens has emerged. These materials do indeed have a property among them which is recognizable, that of electrophilicity in their initial structures. A group of chemicals of this type is shown in figure 1, and it is possible to see that there are two types of electrophilic chemical carcinogenic reactions, enzymatic and non-enzymatic. We will be principally concerned in this discussion with the enzymatic reactions, but both types have cross-linking characteristics.

Those chemicals in the lower part of figure 1 are known to function in the same way as the non-enzymatically active chemicals, but they require an enzymatic transformation to produce the electrophilic reagent which will then attack some of the cellular nucleophiles. One of these chemicals, acetylaminofluorene (AAF), is known to go through the identical sequence of oxidations to give hydroxylamine and then the ester of hydroxylamine (either sulfate or acetate) which gives rise to an electrophile by virtue of the loss of the anion, leaving behind a nitronium ion instead of a carbonium ion. The product of that reaction has been definitely established, and it has rather unique characteristics. The methyl group is oxidized to give, eventually, a hydroxylamine, and the same sequence of events occurs with the acetylaminofluorene. The two nitroso compounds can give rise to methylating diazomethane *in situ*, by virtue of a sequence of oxidations. The diazomethane, in turn, can methylate the various nucleophiles.

The polycyclic aromatic hydrocarbons (PAHs) have remained a mystery for quite a long time because they are really not reactive molecules. It is only in the last decade that the nature of the activation of PAHs has begun to be understood. In general, they are activated by an oxidation mechanism. Much of today's discussion will focus on how the PAHs are activated and what the reactions in the cellular material are as a result of that activation. That particular type of reaction, therefore, will serve as a model for the nature of the chemistry, biochemistry, and biology which are involved in carcinogenesis.

Three different macromolecules which conceivably could be target molecules for any of these electrophilic reactions are shown in figure 1. We have known for some time that if you treat a cell or cell suspension containing all of these molecules with active (activated) materials, they will be covalently linked to all three components: DNA, RNA, and protein. In principle, we do not know which of those targets is the critical one. However, circumstantially, and because of the nature of the biological effects of the chemical carcinogens, it seems almost certain that the important target is the DNA itself since the transformed cells behaves as a mutated cell, i.e. the transformed condition reproduces continuously, which is one of the qualities of tumorogenesis which makes us believe that the DNA is the important target rather than RNA or protein which would not 'remember' these events and continuously reproduce the transformations in the cell.

Having recognized that the product of the activated carcinogen AAF with guanosine has been identified (fig. 2), let us examine that product. The AAF has been acetylated after oxidation, followed by loss of acetate anion, leaving the nitronium behind which can become an electrophile on a collection of DNA bases. It has been shown that the nitronium ion attacks the No. 8 carbon atom of guanine selectively and produces the type of product shown. These experiments, done about half a dozen years ago, were the first case where the electrophilic reaction of a chemical carcinogen with DNA components was firmly established.

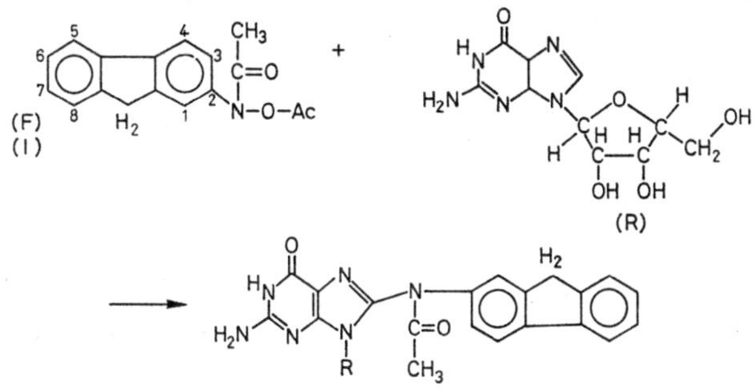

Chemical Carcinogenesis

Fig. 2. Products from activated AAF and guanosine.

It is interesting to note that while the unsubstituted fluorene works and the 7-fluorofluorene works, the 7-iodofluorene does not react with DNA components in this fashion. Two consequences of that fact have been recognized. One is that the hydrogen compounds and the fluorine compounds are both carcinogenic, whereas the iodine compound from the reaction is not carcinogenic. The second consequence resulting from biophysical studies is that AAF and the corresponding fluorine compound affect the helical structure of the DNA whereas the iodo compound does not. The argument is that the carcinogenic compounds can intercalate into the DNA molecule to begin the chemical reactions which follow from it. This is a reasonable, circumstantial argument that the first step is intercalation. Part of the reason that this type of molecule is such an important and selective carcinogen is the fact that it can intercalate in the DNA in a rather special place and perform a rather special reaction in that place. This knowledge has been one of the starting points of our work on chemical carcinogenesis.

About 5 years ago, we began the study of the question of why smog, tobacco smoke and, as it turns out, any combustion product of organic material, contain carcinogens and what the individual carcinogens are. It turns out that the principal carcinogen contained in *all* of the organic combustion products is benzo(a)pyrene (BaP), whose structure is shown in figure 3. To a chemist, it does not seem to make sense that a molecule as non-functional as BaP should be so effective in a biological system. It is a very potent carcinogen and is the principal source of lung cancer, for example, and is present in all organic combustion products. How does the BaP become a reactive material and what are the reaction products?

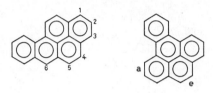

Fig. 3. Structure of BaP (carcinogenic) and BeP (non-carcinogenic).

For the last 15 years, there has been a continuing worldwide study on the reactivity of BaP. (This work has been done mostly in the United States and England, with some effort in France. However, the French study was dominated by the ideas of a theoretical organic chemist to the effect that the important place and reactive position of the BaP is the 4,5-double bond [K region], and the French researchers spent their time and effort trying to show that the reactive product had something to do with these two positions on the molecule.) The conviction of the activity of the K-region epoxide was not so widespread in England and the United States where people recognized that reactivity could be anywhere on this molecule.

The problem has ultimately been solved by examining the BaP metabolites in animals and in animal cells. There are many products of such a chemical carcinogen, and almost every position on the molecule has been oxidized during the course of the metabolism. Which reaction is the important one, or are they all side reactions which really are not important in relation to carcinogenesis?

This is a typical problem created by a mutagen because of the nature of biological systems which are not simple reproducible chemical systems. The dominant reaction products may have very little to do with the biological consequences of the molecule. It may be some trivial step, or side reaction, which is actually the crucial one for the biological consequences of carcinogenesis.

It appears that the 1,3- and 6-positions of BaP are the most active with respect to either oxidation reactions or photochemical reactions. Our first efforts in this area were some photochemical experiments using BaP as one of the chemical reagents with a nucleic acid component (N-methylcytosine) as the other reagent to see if the reaction occurred. It was known that when BaP was painted on the skin of an animal, it would produce skin cancer. It was also known that when illuminated it would produce an even more

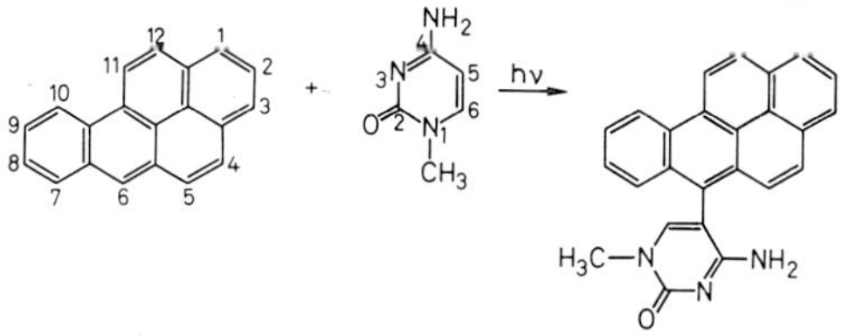

Fig. 4. Photochemical coupling of BaP to N-methylcytosine.

severe skin cancer. The effect of light on this type of a chemical carcinogen is shown in figure 4, and the isolation of the product showed that the No. 5 position acted as an electrophile on the N-methylcytosine. This result also gave us a clue that it would be possible to activate the No. 6 position of BaP. We therefore proposed to activate the No. 6 position of BaP (or the No. 1 or No. 3 positions) by the conjugated bond system which lies between them. This was actually translating photoactivation into chemical activation.

It turned out, however, that this type of reaction is not really important. At various laboratories in the United States and England, experiments were performed to extract the metabolites of BaP and test them to see if they are more potent chemical carcinogens than the original material. On the other hand, we have used BaP, activated it with aryl hydrocarbon hydroxylase (AHH) (an enzyme) to try and couple the BaP with a nucleic acid, or nucleic acid analog, to deduce what has happened in the product. Both approaches have actually converged and have come to the same conclusions within the last year.

AHH is an enzyme which will oxidize the hydrocarbon (BaP, for example) with a cofactor such as reduced pyridine nucleotide and oxygen; the AHH will epoxidize this hydrocarbon in various positions. All of the products of the reaction have been extracted, and it has been shown that this epoxide has been created by a mixed function oxidase which is present in low levels in most mammalian cells. If the mammalian cells are exposed to an aryl hydrocarbon, the enzyme is induced to much higher levels. AHH is an iron enzyme, and a good deal of effort has been devoted to learning how the iron enzyme works. Many things can induce AHH to raise its level in mammalian cells. One of the oxygens of AHH is involved in making an epoxide and the other is involved in making water with a reducing agent. The epoxides can then undergo a hydration reaction which involves opening the epoxide with water to produce a diol (fig. 5). In general, the reaction

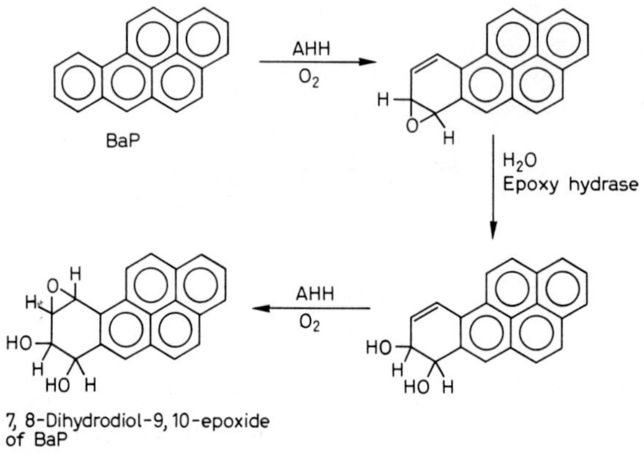

Fig. 5. BaP activation by microsomal oxidases.

occurs as follows: A double bond on the AHH reacts with an oxygen atom to form an epoxide; the epoxide, with water, opens to form the diol. This particular sequence of reactions can be accomplished in most mammalian cells.

The AHH can act on all of the positions in BaP, but none of the derivatives (except the ones shown) turn out to be better carcinogens than the starting material; this is the crucial fact to remember. We have learned this only recently by using a combination of synthetic and enzymatic techniques.

The sequence of events in these transformations seem to be as follows: AHH epoxidizes the 7,8-positions of BaP (or the 9,10-positions). The 7,8-epoxide, upon hydration, gives a 7,8-dihydrodiol; a second epoxidation is even faster (with the first epoxidation it was necessary to epoxidize a partly aromatic double bond). The 9,10-double bond is no longer aromatic and it is very rapidly epoxidized (like styrene) and the 7,8-dihydrodiol does not build up. The 7,8-dihydrodiol-9,10-epoxide is the most potent carcinogenic derivative of BaP. In fact, it is important to consider the stereochemistry of that carcinogen. The one formed enzymatically is more carcinogenic.

This result, which was the combination of metabolic and carcinogenic studies, has focused our attention on the (a) ring as the crucial, active position in the BaP molecule. However, it is still not known what this material is reacting with and how.

Many of the BaP derivatives have been synthesized in our laboratory, but we have not used many of them as stoichiometric reactants with known bits of DNA. We are in the process of doing that type of experiment at the

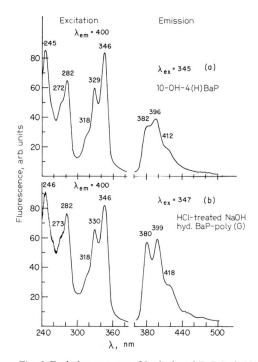

Fig. 6. Emission spectra of hydrolyzed BaP-Poly(G).

moment. This is actually a combination of synthetic organic chemistry and biochemical enzymology. It is known that the diol is a *trans*-diol, but more important is the geometric relation of the epoxide to the 7-hydroxyl. The 9,10-epoxide and the 6-hydroxyl derivatives of BaP are *trans* to each other, which is the most effective carcinogen.

Using AHH from induced rat livers and a variety of DNA analogs as substrates, we found, before we even treated the complex with the AHH, that only guanine-containing polymers would intercalate the BaP to any extent. Calf thymus DNA will intercalate very little; poly(A) will not intercalate the BaP product at all; and the polymeric pyrimidines are not successful in this type of experiment. Poly(G) is the most successful polymeric material for this type of a reaction with chemical carcinogen, but this material is only a model. It will be necessary to use genuine DNA for the final conclusive result to this hypothesis.

The results of optical measurements on the model substances and the products of the reactions of diol epoxide and activated BaP with AHH and with poly(G) and DNA are shown in figures 6 and 7. The enzymatic hydro-

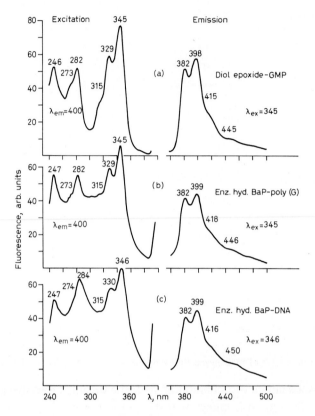

Fig. 7. Fluorescence spectra of BaP covalently linked to nucleic acid.

lysis of the BaP-poly(G) product gives an absorption with two rather sharp peaks which are characteristic of 7,8,8,10-tetrahydrobenzo(a)pyrene in which the benzene ring is completely hydrogenated, leaving a pyrene nucleus; the absorption spectrum is actually characteristic of pyrene. This was the first clue that the product activated BaP with poly(G) was a reaction which had destroyed the 7,8- and 9,10-double bonds of the BaP but had left the pyrene aromatic nucleus intact. There are six positions which apparently were not touched.

The fluorescence spectrum of the product was more critical, showing the fact that the chemically hydrolyzed products were similar. The emission spectra of a tetrahydrobenzpyrene is about the same, but has a different relative intensity. This difference is an important component in gathering our information about the ultimate product. The emission band (380) is less intense than the second (400), whereas in a simple tetrahydrobenzpyrene (7,8,9,10-tetrahydrobenzpyrene) the first band is always the most intense in fluorescence. There is one model case, however, in which the first fluorescence (emission) band is less intense than the second, and this occurs with 10-hydroxy-tetrahydrobenzpyrene. In the No. 10-position there is a substitution other than hydrogen. This example is the only one which shows the first fluorescence emission less intense than the others. That fact indicates that there will ultimately be a tetrahydrobenzpyrene product, but it will also have a substituent on the No. 10-position, probably bearing an unshared electron pair.

The fluorescence information tells us that the guanine is on the No. 9- or No. 10-position, with two hydroxyls on the 7- and 8-positions of the BaP ring. When you examine the structure of the 7,8-diol-9,10-epoxide of BaP, you can begin to surmise what the reaction product really is. When the 7,8-diol-9,10-epoxide reacts with the guanine of the poly(G), the obvious place for the epoxide to open is such as to put the carbonium conjugate with the pyrene. This is a stable carbonium ion and we then have an electrophilic reagent. So far, the only model we have is that of the 8-position of the guanine as the electron-rich position, as evidenced by its reactions with AAF shown earlier. I have therefore surmised that one possible product of the reaction between the activated and intercalated BaP is an electrophilic attack on the No. 8 carbon atom of guanine. We could then remove the C-8 proton to the epoxy-oxygen, insert the double bond again, with the resulting product being 10-(guanyl)-7,8,9-trihydroxytetrahydrobenzo(a)pyrene. Alternatively, we could ring-close to give a ring-closed product, because when we treat the ultimate reaction product with toluene sulfonic acid, we do not find a new double bond: If there were a hydroxyl at C-9 and a hydrogen at C-10, we should get a double bond upon treatment with toluene sulfonic acid. An alternative point of C-10 carbonium ion attack would be on the exocyclic nitrogen atom of the guanine, giving a simple C-10 N-substituted C-9 hydroxylated 7,8,9,10-tetrahydrobenzpyrene derivative.

Biological Consequences of Chemical Carcinogenesis

It is now necessary to introduce the concepts that are current in the chemical and viral cancer community in discussing the biological consequences of chemical carcinogenesis. The process of viral transformation is

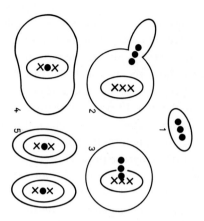

Fig. 8. Transforming infection with SV-40 virus.

shown diagrammatically in figure 8. Here the black spots represent pieces of DNA containing information which, when integrated into the chromosomal DNA of the cell, will transform it into a tumor cell. If the virus simply infects, the replication of the virus lyses the cell, and the cell is not transformed. However, some part of the viral genome is integrated into the cellular genome, and the cell can be transformed into a tumor cell. If a cell is transformed, such cells overgrow each other, creating foci (individual cells piled on top of each other). Cell transformation by virus can be assayed by focus-formation, indicating the degree of viral transformation of the original cell culture.

The insertion of the oncogenic information, or the whole viral genome (or some crucial part) which contains the oncogenic information into the chromosomal DNA, occurs through a mechanism as yet unknown. The scheme for cell transformation, including chemical function, is shown in figure 9. Here it is seen a DNA virus gets inside the cell through the function of various nucleotide hydrolyzing enzymes (endonucleases, exonucleases, ligases) which can insert bits of DNA into the chromosomal DNA. The result is chromosomal DNA containing some combination of the viral DNA which gives rise to the transformed cell. (For an RNA virus it is necessary to go through a special enzyme, RNA-dependent DNA polymerase [RDP] to make a DNA copy of that RNA virus and insert that. Both viruses act by inserting bits of information representing the viral oncogenic information into the chromosome of the cell.)

Chemical Carcinogenesis

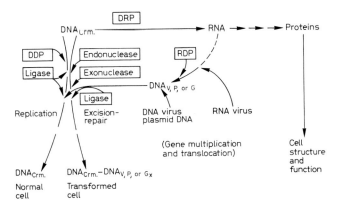

Fig. 9. Scheme for cell transformation, including chemical function.

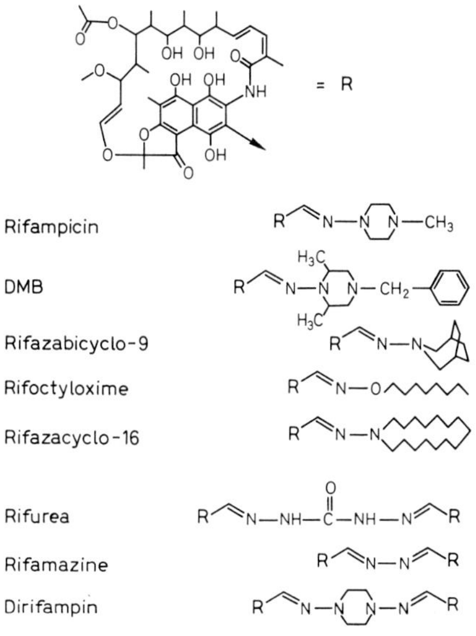

Fig. 10. Some synthetic modifications of rifamycin.

If we had an RNA virus and it had to go through RDP to get to DNA to be inserted and transformed, and if we could specifically block this enzyme, we could then prevent cell transformation by that virus. This experiment was done several years ago because there was a drug available which would inhibit that particular reverse transcriptase enzyme. We were able to do that

Table I. Effect of DMB on transformation of UCI-B by MLV

DMB μg/ml	UCI-B foci/10^5			XC plaques/10^5			XC plaques/5×10^5		
	1	2	av.	1	2	av.	1	2	av.
0	120	158	139	272	231	251[1]	224	242	233
3	108	71	89	266	216	241[1]	308	282	295
6	70	52	61	316	356	336	258	228	243
9	48	36	42	249	322	285	326	306	316

[1] Some plaques contain foci.

particular type of experiment, using synthetic modifications of the drug, rifamycin (the chemical formulas of which are given in fig. 10), and information concerning the effect of these drugs on cell transformation is given in table I.

How is this work related to chemical carcinogenesis? I had the idea that the chemical triggers integration of some endogenous information which is in the cell and which is not being expressed. In order to test this hypothesis, we tried to find a system where the chemical alone appears to produce a tumor. We knew, for example, that there were certain strains of rats where a single injection of a chemical carcinogen such as dimethylbenzanthracene (DMBA) produced mammary tumors in 9 weeks, killing all the animals; this result occurred each time the experiment was performed. I felt that this result indicated that this particular strain of animals carries endogenous oncogenic information which is triggered by the chemical carcinogen. If it was an RNA virus, the rifamycin should prevent or slow down the carcinogenesis, which actually is the case. The results of an experiment of this type are shown in figure 11.

The converse of this former experiment has been done recently using a strain of rats which does not produce a tumor at all with the carcinogen alone. However, if these rats are given an adenovirus at a suitable time before administration of the chemical carcinogen, then there is tumor formation. Adenovirus alone has a certain rate of tumor production, the chemical alone produces no tumors, but the chemical carcinogen with the adenovirus is much more than additive. This experiment is the opposite of the first type where the animal itself carried the oncogenic information which was not being expressed, presumably in the form of extrachromosomal information or in the form of a putative virus. In the experiments which I have just de-

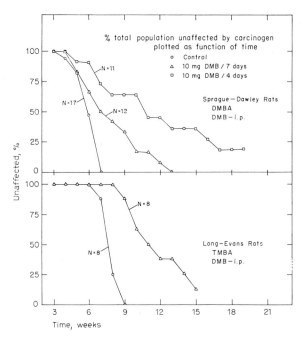

Fig. 11. Prophylactic effect of DMB against DMBA and TMBA in rats.

scribed, there is a strain of animal which does not respond to the chemical treatment unless it also receives the viral infection.

These same types of experiments can be done in tissue culture where the results are somewhat more reliable than the information obtained from whole animal experiments. The viral transformation of hamster cells following treatment by a chemical carcinogen (4-nitroquinoline oxide, 4-NQNO) has been studied, as an example. The adenovirus alone without the chemical treatment with 4-NQNO produces very few foci; with 4-NQNO treatment and virus, the number of foci increases; eventually, the chemical effect on the repair mechanism of the cell is over, and the excess integration ceases. This result indicates a synergism between the chemical and viral transformation, as indicated in table II.

How can this synergism be understood? The chemical starts a process of manipulation of the cellular DNA to repair on; the repair enzymes are operating to replace the defective DNA. If there is a source of misinformation (oncogenic information) in the cell at the same time, there is a certain probability that the oncogenic information will be integrated into the cellular genome during that repair operation. It turns out that there are many dif-

Table II. Transformation of Syrian hamster cells by SA-7 following 4-NQNO treatment[a]

SA-7 introduced: time after 4-NQNO treatment[b], h	Number of foci/2 × 10⁶ plated cells	
	SA-7 only	4-NQNO (2×10^{-6}M)-1.5 h + SA-7
0 hrs.	17	435
12	17	26
24	17	12
48	17	7

[a] From STICH, H.F.; in *Topics in Chemical Carcinogenesis*, p. 23 (University Park Press, Baltimore 1972).
[b] Unscheduled DNA synthesis induced by 4-NQNO is over in about 8–10 h.

ferent types of nucleic acid repair enzymes, each one specific for a different type of error, and each one, in some cases, for different base sequence. It is only relatively recently that the variety of clipping enzymes (endonucleases and exonucleases) have become visible in reactions of this type.

I want to suggest that the chemical (carcinogen) produces a distortion of some kind in the DNA which, as a result of that distortion, is subject to the attempt to replace the distorted piece by the variety of repair and replication enzymes which are present. If oncogenic information is present, it has a probability to be inserted. The chemical thus enhances the probability of insertion of oncogenic information. The chemical alone is a mutagen, but is not a carcinogen by itself. The carcinogenic result is due to the presence of some other piece of information which the chemical triggers to insert into the cell.

These speculations which I have just discussed are shown diagrammatically in figure 12. The chemical carcinogen puts a 'kink' of some kind in the DNA to start the whole process, leading to an accelerated probability of insertion of the oncogenic information which is there from some other source. You have heard comments to the effect that the test for mutagenicity is a good enough test for carcinogenicity as well, but I do not believe this is the case. All carcinogens are mutagens to be sure, but all mutagens may not be carcinogens because the carcinogenic information is not there. Mutagenesis generally leads to a lethal event. I believe that a single point mutation, which is what the chemicals can induce, cannot introduce enough in-

Chemical Carcinogenesis

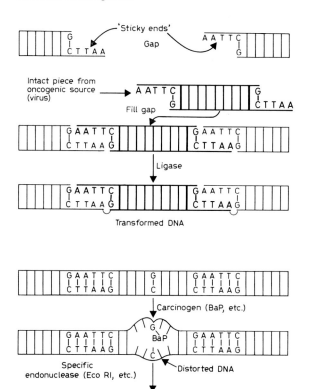

Fig. 12. Schematic proposal for collaboration of chemical carcinogen and oncogenic information (parts a and b).

formation to lead to transformation. The introduction of that much information really means the introduction of a large piece of DNA, or the removal of a large piece of DNA.

Using what we know about the insertion of known bits of DNA which can be done in the test tube (I am presuming that the same kinds of events go on in the cell) from one species of DNA, we can insert them into another species of DNA using suitable clipping enzymes (exonucleases, endonucleases, etc.). When the chemical carcinogen forms a covalent complex with guanine, the double helix must be completely destroyed in that vicinity, warping the geometry of the molecule. BaP has been completely intercalated, and when covalent linkages are formed and the guanine is pushed out of the double helix, the helical structure is distorted around the point where the carcinogen has formed the covalent link with a suitable base. The distortion will be recognized by a variety of enzymes which must act upon it by some mechanism. The specificity of those enzymes is not yet known. It is also not known whether the enzymes must be induced, or whether the enzymes are always present, and a clonal selection process operates similar to that in which the antigen induces new cells to make a particular antibody.

In any case, some event occurs which leads to the beginning of the removal of this distortion by a restriction enzyme. This may lead to insertion, leaving a gap behind with sticky ends (characteristic pieces of DNA information). The same clipping enzyme (endonuclease, etc.) acting on exogenous information, or at least information which is not chromosomal in the cell, will produce other such sites with complementary ends. It can thus produce a piece from an oncogenic virus with corresponding sticky ends which, in turn, can fill the gap. The breaks can be 'sewn up' with the ligase, resulting in the transformed DNA with the oncogenic information inserted. This is not a real 'repair' mechanism, at least it is not conceived of in that fashion. How the actual repair mechanism may work in this case, I do not know. The diagram gives only one concrete example of how information which is not in the chromosome could be taken and inserted into the chromosomal DNA as a result of a chemical action by a particular type of a chemical carcinogen or chemical mutagen.

Most recently in our efforts to define and measure the differences between the normal and transformed cell, we have explored a new method of probing the characteristics of the cell surface, the most easily accessible part of the cell for possible treatment. This new method has not only yielded information on the differences between normal and transformed cells, but also seems very likely to provide a much better, simpler and more quantitative method for measuring the rate of appearance of transformed cells in a cell population.

It is clear that these two objectives – (a) to define the differences between the normal and transformed cell surface, and (b) to be able to measure the kinetics of transformation using the degree of that difference – are both extremely useful and important. The latter one will, of course, allow us to

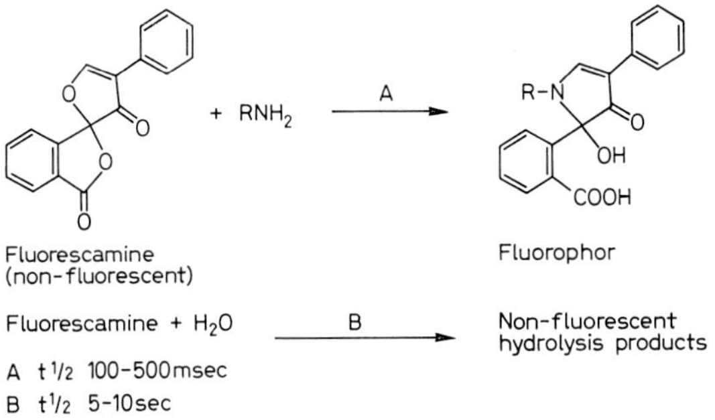

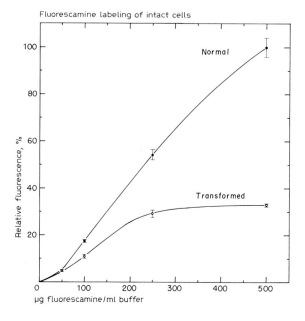

Fig. 13. Reaction of fluorescamine with primary amines and concomitant hydrolysis of reagent.

Fig. 14. Fluorescamine labeling of intact cells.

explore more readily that synergism for which we have only the focus assay to guide us.

The method basically depends upon the availability of a chemical reagent which will react rather specifically with free amino groups, either ter-

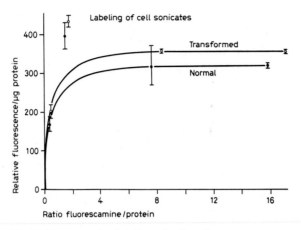

Fig. 15. Fluorescamine labeling of cell sonicates.

minal ends of proteins or other biological molecules, to form a fluorescent product. Neither the reagent itself nor its hydrolysis product is fluorescent. The reagent is not transported across the cell membrane in a reactive form. Therefore, when the cell is treated with this reagent, only the cell surface amino groups become fluorescently visible.

The fluorescent reagent and its reaction are shown in figure 13, and the results of treating normal and transformed cells with this reagent are shown in figure 14. Here it is clear that the available amino groups on the cell surface for labeling are somehow reduced by transformation. If the cell membrane is broken and the entire protein population of the cell (both internal and external) is allowed to make contact with the reagent, the difference between the two cell populations is very little, which is shown in figure 15.

An attempt to determine the nature of the particular proteins which have been deleted from the cell surface upon transformation is shown in the electrophoretogram of figure 16. One of the potentially fluorescent proteins which is absent from the transformed cell membrane is clearly shown as the missing band in this figure. It seems very likely that this missing protein in the transformed cell membrane is identical with the large external transformation-sensitive (LETS) protein which has been described by other methods and whose detailed character and function are yet to be determined.

Finally, we have used two fluorescent stains on the same group of cells. The first is propidium iodide which enters the cell and stains the nucleic acid, the fluorescence intensity of which is a measure of the amount of

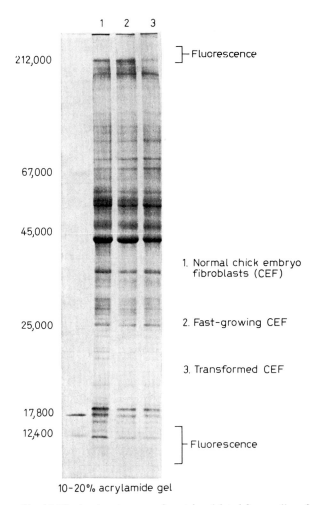

Fig. 16. Electrophoretogram of proteins deleted from cell surface upon transformation.

nucleic acid present in the cell. The other stain, fluorescamine, is used to label the cell surface. These doubly labeled fluorescent cells are shown in figure 17, and with such doubly labeled cells and a flow microfluorometer (an instrument for measuring the fluorescence intensity for individual cells) which can be set to measure at least two different colors of fluorescence, we have been able to distinguish very clearly a population of normal cells from a population of transformed cells. The data showing this capability are ex-

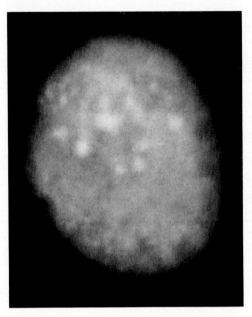

Fig. 17. Doubly labeled Balb/3T3 cells (labeled with propidium iodide and fluorescamine).

hibited in figure 18 in which particular populations of each of the cell types are shown in the top two panels. It is clear that the ratio of surface fluorescence to nucleic acid fluorescence is very much smaller for the transformed cells than for the normal cells. From our analysis, the population shown in the various panels of figure 18 contain transformed cells as follows: (a) 4.5% transformed cells; (b) 34.6% transformed cells, and (c) 82.6% transformed cells.

With this information, we are now prepared not only to explore the nature of this characteristic surface protein which is absent from the transformed cells, but also to use this quantitative characteristic of a cell population to measure the carcinogenicity of chemicals by a quantitative determination of their synergisms with suitably transforming virus in a selected cell population.

As this work proceeds, the more precise relationship between a chemical's ability to induce mutations and its ability to induce malignancy can be defined. In fact, it is very likely that this kind of assay for carcinogens will be the quickest and most relevant one for examining chemicals in our environment.

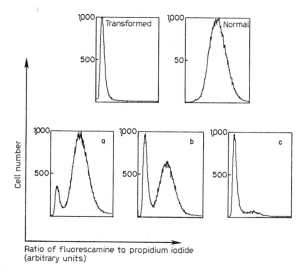

Fig. 18. Detection of transformed cells using fluorescent probes on MSV-infected Balb/3T3 cells.

If our synergistic proposal withstands these tests, the obvious confirmatory experiment will be to demonstrate the integration of the oncogenic information from some external viral source induced by the chemical. Plans are under way to accomplish this.

Summary

The first step in the generation of a malignancy seems to be a transformation in the genetic apparatus of a single cell. The ultimate nature of the cancer which appears is a result of the interaction of that change with the control and regulatory apparatus of the whole animal. It appears that the primary cellular change which may be induced by physical, chemical or biological agents (or a combination of them) may be something which is common to all carcinogenesis. The nature of that primary change and how it may result from the action of viruses, chemicals, and radiation (or interaction between them) is the subject of this discussion.

Dr. M. CALVIN, Laboratory of Chemical Biodynamics, University of California, *Berkeley, CA 94720 (USA)*

The Impact of Nuclear Technology on the Natural Environment and Human Life

RAYMOND LATARJET

Institut du Radium, Paris

Within its narrow limits, this presentation will focus on the most fundamental aspect of this immense problem, that of a diffuse and permanent pollution by small amounts of ionizing radiations. Unlike most chemical pollutions, this has the peculiarity of being superimposed on a natural background, whose level is well known and remains almost constant, to which the biosphere is well adapted, and which is likely to have played – and to still play – a role in the evolution of species.

I. Dispensable and Indispensable Pollutants

When the former cause a problem, it is easy to get rid of them. The use of dimethylaminoazobenzene (butter yellow) was forbidden, once it was found that this compound is a powerful hepatic carcinogen. A pollutant can be considered as indispensable when its advantages obviously prevail upon its dangers. Such is the typical case of ionizing radiations in radiotherapy. One cures a cancer with radiations which themselves can produce a cancer. But the probability of curing the cancer which is irradiated is much higher than that of eliciting another cancer. Similarly, in radiodiagnostic, X-rays are more beneficial than harmful (except perhaps in the indiscriminate systematic screenings). One might say that another typical indispensable pollutant is tobacco, since, despite all warnings, demand exceeds fear. The smoker goes on with smoking although he is told that his life span is likely to be shortened by 10 years (which, by the way, is not bad, on the demographic point of view).

In the face of an indispensable pollutant, the problem resides in reducing its harmful effects, while keeping the advantages of the system which produces it. This problem can be schematized according to figure 1 in which are drawn: (a) the curve of harmful effects as a function of the dose (whose

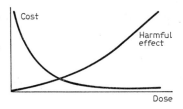

Fig. 1

shape depends on the agent); (b) the curve giving the cost of protection against the agent, as a function of the dose.

The harm increases with the dose. The cost of the protection rapidly increases when the dose decreases. It is rather easy to eliminate the excess. But elimination becomes very difficult and expensive for the last traces. The sensible attitude consists in finding the best compromise between: (a) excessive risks (which sometimes are hidden or minimized); (b) unnecessary protection.

It is not realistic to systematically aim for additional protection. This attitude, which sometimes looks like a 'demagogy of pity' would be admissible if it were equally aimed at all harms and risks, which is never the case.

Are radiations disseminated by nuclear reactors an indispensable pollutant? Indisputably, thanks to their mutagenic and lethal activities, they are a risk, and we cannot produce atomic energy without disseminating some radiations – whether as radioactive nucleides or electromagnetic radiations. In other words, is nuclear energy indispensable? My answer is 'Yes – for the few decades ahead'.

In western Europe[1], in 1987, without nuclear energy, the deficit in the energy required to keep the present standard of living and the present level of industrial activity (in zero-growth conditions) would amount to 20–25%. The sociological harm of such a deficit would obviously exceed those of the radiation hazards as they are already controlled. Therefore, we are compelled to develop a nuclear energy program; and this conclusion is valid outside western Europe. But, in my opinion, this situation will not last. For many reasons which cannot be discussed here, the energy of the future will be that of the past, the eternal and non-polluting energy from the sun. The total amount of energy used by man on Earth is presently about 10^{-4} that of the solar energy which falls on our planet. That which is used in the United States is only about 1/700 that from the sun which falls on this country. To me, the future belongs to biological conversion of solar energy through hydrogen-producing photosynthetic processes. And if I had to bet, with all

---

[1] From the present estimations of the European Community Advisory Committee.

the required caution and modesty, I would say that within less than half a century the problem will be solved. Consequently, the development of nuclear energy should be carried out, taking into account the short duration of the need.

Nevertheless, since we need it for one or two generations, the present problem is relevant.

II. Radiation

Radiations belong to the class of pollutants which attack the genetic material. Thus, they are endowed with three main biological activities: they are lethal for the cells, they are mutagenic, and they are carcinogenic. The first of these activities is the most important one in view of its consequences, but it is to the third one that public opinion is most sensitive – very naturally, but wrongly. The lethal effect at the level of monocellular organisms owes its importance to the fact that these organisms (bacteria, algaes, yeasts, molds, etc.) are the only ones capable of synthesizing the amino acids which are the building blocks of proteins. In order to manufacture their proteins, the higher organisms must feed themselves with amino acids produced by the microorganisms. If one sterilizes the microflora, the biological chain progressively disappears up to man at the top.

A natural example is given by the case of Rio Negro, in Brasil, a large river which joins the Amazon at Manaus, and whose very pure water has always fed that town without being purified. There are no fish in Rio Negro, and therefore man has never settled on its banks, whereas houses are numerous all along the nearby Amazon which is very rich in all sorts of living species. Prof. L.R. CALDAS, from Rio de Janeiro, and his team, have recently discovered that the land around Rio Negro contains a bacterium which produces a powerful photodynamic pigment, 'violacein' which the rain carries into the water of the river. Through the action of sunlight sensitized by violacein, the microflora of the river is destroyed. No higher organism can grow in those sterile waters.

Moreover, it should not be forgotten that at the surface of the oceans, the microflora constitutes the most important sector of photosynthesis on earth, that is the most important for the regeneration of oxygen and for the production of nutritive matter.

Therefore, when one considers the hazards of pollution, it is childish to worry mostly about the direct dangers to man, whereas the integrity of the microflora should come first in our preoccupations. To sterilize the sea is more serious for life in general and for man in particular than to destroy towns.

The present permissible doses of radiations have been fixed in order to safeguard the genetic equilibrium of human populations. A limiting dose was chosen which approximately doubles the natural radiation at ground level, i.e. about 1/20th the dose which is considered as doubling the mutation frequency in man.

It is usually easy to determine, by direct experimentation, the dose-effect curves for radiation-induced mutations and lethality in microorganisms. In general, these microorganisms are more resistant to both phenomena than the cells of higher organisms, namely mammals. The ratio of sensitivities is usually higher than 10. Moreover, thanks to the short generation time and to the high number of individuals, micro-organisms display a remarkable adaptation (by the way of mutation-selection) to increases in the level of surrounding radiation. For all those reasons, one may conclude that the doses presently admitted as permissible are harmless for the microflora. In fact, the latter is much more threatened by chemical pollution than by radiation.

III. Pollution

What we call pollution is in fact the spread of a harmful agent at low doses over a high number of individuals. This situation is easily open to experimentation with microorganisms, when numerous individuals can be treated in a homogeneous and well-controlled fashion. But it raises difficult and complex problems when one wants to operate on higher organisms, especially on mammals. The so-called mega-experiments are rather frustrating. When one increases the number of treated animals in order to increase the number of significant alterations, one increases the heterogeneity of the experimental parameters and, consequently the size of noncontrolled fluctuations. For example, let us consider an experiment with so weak a dose that only one positive result can be expected out of 10^4 animals. If one wants to count ten results, which is a minimum, one must operate with 10^5 animals (plus the controls). It is impossible to operate with 10^5 mice – the most convenient mammal – under satisfactory conditions. The parameters will vary: strain, food, age, housing, technician, etc. I doubt that the famous experiments carried out by Mr. and Mrs. RUSSELL at Oak Ridge between 1950 and 1960 will ever be repeated.

There are two possible ways of obtaining quantitative information on the effects of low doses of a pollutant on man: (a) epidemiological studies when the conditions can yield significant results; (b) extrapolation from the results obtained with lower organisms.

An example of a is the statistics of cancers among the survivors at Hiroshima and Nagasaki. It covers several hundreds of thousand individuals. It required a gigantic study carried out for 25 years by the University and by an Institute specially created for that purpose. Another example is given by the statistics of hemopathies among the populations which live at high altitudes in the Andes, where the cosmic radiation is about three times as high as at sea level.

A recent attempt at extrapolation deals with the frequency of gene mutations induced by low doses of ionizing radiations. ABRAHAMSON and his colleagues claimed in 1973 that, when one compares the efficiency of radiation for the induction of such mutations in different animal and vegetal species, there is a linear relationship between that efficiency, i.e. the number of mutations per genetic locus and per rad, and the amount of deoxyribonucleic acid (DNA) in the haploid genome of the cell of the species. According to the amount of its DNA, which is well known, man is located between mouse, barley, and tomato. Thus, one can deduce from the curve that 1 rad will induce in man 3.5×10^{-7} mutation per locus.

If true, this so-called ABCW relationship (from the initials of the authors) would be of great help for our knowledge of what happens in man. Its validity has been substantiated by the finding that a similar relationship seems to hold true in the case of a chemical mutagen, an alkylating agent, ethyl methane sulfonate. However, the ABCW relationship is surprising. In order to interpret it, one must admit either (a) that the entire genome participates in the mutation of a locus, or (b) that when the amount of DNA increases, the number of genes remains the same, but their individual size increases accordingly. These two assumptions are not easy to admit. Numerous facts argue against them. Thus, the ABCW relationship is being criticized, and the easy extrapolation from other species to man remains open to discussion.

IV. The Threshold

When low doses of an agent act on a biological system, one says that the effect has a threshold when the dose-effect curve does not start from the origin, i.e. that a threshold dose exists below which the agent does not produce any effect. Conversely, one says that there is no threshold when a dose, however low it may be, has a non-nil probability of producing the effect.

This question is important because it governs, among others, the notions of risk and of responsibility.

We know some radiation effects which have no threshold. Such are gene and chromosome mutations; such is lysogenic induction in bacteria. For somatic effects in higher organisms, the question is not solved. As already mentioned, direct experimentation is seldom, if ever, possible, and this is particularly true for cancer in man.

It has been wrongly argued that radiation carcinogenesis in man is a non-threshold phenomenon. The rational of this argument is the following: 'The transformation of a normal cell into a cancerous one is a mutation (per definition). What is true of mutations is also true of cancerization. In particular, since there is no threshold for mutations, there is also none for cancer. If the frequency of gene mutations is extrapolated to cancer, then natural radiation would be responsible for a noticeable fraction of the so-called spontaneous cancers, and any increase in the natural level of radiation would be responsible for new cancers.'

In this reasoning there is a confusion between the malignant transformation of a cell, which is a mutation, and the clonal emergence of that transformed cell into a tumor. The presence of a transformed cell is not a cancer, and this is fortunate because all of us probably harbor numerous cancer cells, whereas only one out of five among us will develop a cancer. The cancer will appear only if a transformed cell finds the conditions which are suitable for its growth, in particular if it takes over the immunological reactions from the host. It is at that level that phenomena might intervene which establish a threshold. In fact, we know several experimental systems in which carcinogenesis by radiation displays a high level threshold. Such are, for example, lymphoid leukemia induced by X-rays in mice, and skin cancer induced by ultraviolet radiation in the same species.

In the absence of any direct demonstration of the existence of indisputable thresholds in the radiation somatic effects in man, I think it useful to introduce the concept of 'practical threshold'. There is a natural basic frequency of those effects. Let us say that, in the absence of pollution by radiation, there is a natural background for them. An excess of newly produced radiation will intervene by increasing the frequency. I define the 'practical threshold' as the dose of radiation below which no significant increase of the background frequency is detected.

For example, natural ionizing radiations are represented by cosmic rays and by those from the radioactivity of the ground. The amount of cosmic radiation triples between sea level (35–40 mrad per year) and the altitude of

3,000 m (90–130 mrad per year). In France, the doses received by the gonad of men (taking into account the absorption by the superficial tissues) varies from 45 to 90 mrem per year in the plains to 180–350 mrem per year in the granitic regions (because granit contains a small amount of radium), i.e. a 4-fold ratio. Numerous statistical studies carried out in several countries have failed to find significant differences in the frequencies of somatic alterations among the populations of those regions. Consequently, doses in the range of 100–350 mrem per year appear to be below the practical threshold, and it seems unnecessary to aim at further protection or limitation (in this respect, recent decisions taken in the United States seem excessive to me, and I doubt that they will be respected in the future).

The International Commission on Radiation Protection has recommended that the permissible doses of ionizing radiations spread over large populations should remain within a range which would double the natural radiation. This recommendation is usually followed by all responsible authorities throughout the world. We have just seen that these doses are well below what I have called a practical threshold for somatic alterations in man. As regards the long-term genetic changes and consequences, nothing is quite certain. The dose of radiation which doubles the frequency of mutations in man cannot be directly measured, as already pointed out. It varies with the conditions of irradiation, in particular with the time of exposure. However, we can estimate an average value of about 60 rad, i.e. about 20 times the amount of natural radiation accumulated over a period of 30 years – which is the average life span of the human gametes involved in reproduction.

I consider that to allow, as a maximum in wide radiation pollution, about 1/20th the dose which would double the frequency of natural mutations, is a wise and realistic compromise between the conflicting needs expressed in figure 1. This dose is likely to impose a very light genetic load on the exposed population, a load which it will be very difficult, and perhaps impossible to detect, a load which, in any case, will remain well below that which we owe to modern medicine. By preventing the early natural elimination of genetically altered individuals, the latter allows them to reproduce and to expand their defects before they themselves disappear.

In addition, we should not forget that radiation pollution is only one slight part of the mutagenic pollution which we are exposed to. Today, chemical mutagens take a much greater part. One cannot correctly treat a minor danger while ignoring the major one. And this brings me to end this presentation with a few brief considerations about a new concept, that of radiation equivalence for chemical mutagens.

V. Chemical Pollution

The general concern about radiation pollution arose shortly after world war II because of the horrors of Hiroshima and Nagasaki, and because of the foreseeable development of nuclear energy for peaceful purposes. As early as 1954 ICRP enacted its first recommendations about permissible doses of radiations for most circumstances of utilization and exposure. Today, more than 25 years later, no similar guidelines exist in the field of chemical mutagenic pollutants, although there is a great need for them. For example, we would wish to quantitatively compare chemical pollution by coal- and oil-fired power stations, and radiation pollution by the nuclear power plants producing the same amount of energy. We would like to have a similar comparison between urban pollution by polycyclic hydrocarbons and radiation pollution with regard to carcinogenic effects in man.

It is well known that many chemicals, including common pollutants, can exert effects similar to those of ionizing radiations. The lesions which they produce in the genetic material are very similar. The similarity is such that a cellular strain which is hypersensitive to radiation, because it lacks some repair system for the lesions in DNA, is also hypersensitive to most chemical mutagens. This suggests that one can establish an equivalence between the 'dose' of a chemical and a dose of radiation on the basis of the effects produced on some biological system of reference. Thus, 'current limits for radiation exposure could be regarded as a starting point for the control of chemicals having similar genotoxic effects'.[2]

This is a new concept which presents many unsolved problems, and is therefore open to criticism. Conversely, mutagenic chemical pollution cannot remain in its present anarchic situation, while radiation is well handled and controlled. Since radiations have shown the way which must be followed, and since they are a minor component in the overall mutagenic pollution, there is a great need for rad-equivalences. We are undertaking a great endeavor whose goal is to bring order in the field of mutagenic pollution.

[2]Report given by consultants to the International Agency of Atomic Energy, Brighton, 1976.

Dr. R. LATARJET, Fondation Curie, Institut du Radium, Section de Biologie, 26, rue d'Ulm, *F-75005 Paris* (France)

Epidemiological Research on the Relationship between Tobacco, Alcohol and Cancer

ROBERT FLAMANT

Département de Statistique Médicale de l'Institut Gustave Roussy et
Unité de Recherches Statistiques de l'INSERM, Villejuif

A lot of papers have been published in this field, especially on the relationship between tobacco and cancer. Our purpose is to do a critical review of the major epidemiological features, to assist those who are in charge of promoting a policy of prevention.

Relationship between Tobacco and Cancer

Cancers Associated with Smoking

Table I shows the cancer sites induced by smoking determined from some important prospective studies. In order of decreasing risk, they are: lung, larynx, pharynx, oral cavity, oesophagus and bladder.

Table II shows at the other hand the other cancers apparently non-related to the use of tobacco.

Cancer Risk with Respect to the Method of Smoking

The effect of the method of smoking is shown on table III. These results are drawn from a French case-control study of nearly 4,000 cancer cases [1]. All the sites are related to the amount smoked except the case of oesophageal cancer. Three sites, lung, larynx and bladder, are found to be related to cigarette smoking only and inhalation of the smoke.

Although these characteristics of smokers are associated with each other, it has been established that each one of them has had its proper effect. On the other hand, the protecting effect of using filters has been studied. Table IV shows that the risk of developing a carcinoma of the lung is lower in men who have smoked filter-tipped cigarettes for at least 10 years than

Table I. Cancer mortality in smokers relative to non-smokers[1]: cancer induced by smoking

Type of cancer	Current smokers of cigarettes only				Current smokers of pipes and/or cigars	
	US veterans	US 9 counties	Canadian veterans	British doctors	US veterans	British doctors
Oral cavity	4.1	2.8	3.9	–	3.9	–
Pharynx	12.5					
Oesophagus	6.2	6.6	3.3	3.0	4.1	2.2
Larynx	10.0	13.1	–	–	7.3	–
Lung	12.1	10.0	11.7	16.9	1.7	4.6
Bladder	2.2	2.4	1.7	2.1	1.1	0.5

[1] Observed deaths divided by expected deaths at non-smokers' rates. After DOLL [10].

Table II. Cancer mortality in smokers relative to non-smokers[1]: other cancers

Type of cancer	Current smokers of cigarettes only				Current smokers of pipes and/or cigars	
	US veterans	US 9 counties	Canadian veterans	British doctors	US veterans	British doctors
Stomach	1.6	2.3	1.9	1.5	1.2	1.3
Intestines	1.3	0.5	1.4	1.4	1.2	1.3
Rectum	1.0	0.8	0.6		1.1	
Pancreas	1.8	–	–	1.4	1.1	1.0
Prostate	1.8	1.6	1.5	0.7	1.4	0.6
Kidney	1.5	1.5	1.4	–	1.2	–
Lymphoma	1.3	–	–	–	0.8	–
Leukaemia	1.4	–	–	–	1.2	–
All others	1.4	1.7	1.4	0.9	1.1	0.6

[1] Observed deaths divided by expected deaths at non-smoker's rates. After DOLL [10].

Table III. Comparison of cancer at different sites in cancer patients and controls[1]

Site		Number of cases	Percent smokers	Total amount smoked by smokers (cigarettes/day)	Percent cigarette-only smokers among smokers	among cigarette smokers	Percent of subjects inhaling cigarette smoke among cigarette smokers
Lips	cancer	49	92 +	16.6	91	95	26
	controls	49	71 +	12.1 +	83	88	30
Tongue	cancer	164	95 +	16.0	86	92	53
	controls	164	81 ±	13.7 +	83	90	38 +
Oral cavity	cancer	144	97 +	18.1 +	86	92	48
(other sites)	controls	144	88 +	14.6 +	88	90	46
Oral	cancer	141	99 ±	19.6 +	86	90	47
mesopharynx	controls	141	77 ±	15.4 +	92	93	45
Hypopharynx	cancer	206	96 +	19.1 ±	82	86	48
	controls	206	88 +	15.1 ±	83	87	42
Oesophagus	cancer	362	97 ±	16.8	82	87	39
	controls	362	83 ±	16.0	85	90	38
Lung	cancer	1,159	96 ±	20.1 ±	93 ±	94 ±	58 ±
	controls	1,159	80 +	15.6 ±	84 ±	89 ±	41 ±
Larynx	cancer	249	98 ±	19.7 ±	91 +	93	60 ±
	controls	249	82 ±	15.6 ±	83 +	90	40 ±
Bladder	cancer	214	89 +	16.9 +	93	94	54 +
	controls	214	80 +	14.4 +	88	91	37 +

[1] After INSERM [1].

Table IV. Risk of developing squamous or oat cell carcinomas of the lung in men who continue to smoke, by type of cigarette

Type of cigarettes smoked	Risk in smokers divided by risk in non-smokers, by amount smoked: Number of cigarettes smoked per day			
	1–9	10–20	21–40	41 or more
Filter-tipped cigarettes for at least 10 years	5	14	23	105
Plain cigarettes	20	23	42	169

After Doll [10].

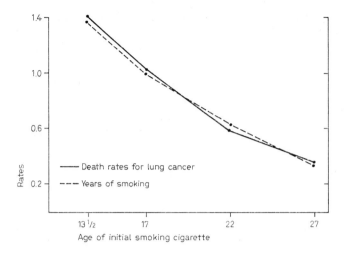

Fig. 1. Lung cancer death rates in relation to ages of initial smoking. After TODD [2].

those who have smoked plain cigarettes, even taking into account the amount smoked.

Cancer Risk with Respect to the Duration Period of Smoking

From DORN's study of US Veterans, it has been shown that there is a close numerical association between the lung cancer death rates for different ages of initial smoking and the duration of smoking raised to the fourth power [2].

This is shown in figure 1. To facilitate comparison, the lung cancer death rates for age of initial smoking have been expressed as ratios, taking the rate for initial smoking at age 17 as 1. The fourth powers of the duration of smoking have similarly been expressed as ratios, taking the duration from age 17 as 1.

Next, the effect of giving up smoking is a very important point to be considered in view of preventability of cancer. Table V indicates that lung cancer death rates are lower among ex-cigarette smokers than those who continue to smoke and even lower when they have stopped for a longer period of time [3]. One interpretation is that when a long-term cigarette smoker gives up smoking, the risk of lung cancer gradually diminishes and eventually approaches the risk for people who have never smoked. But a

Table V. Age-standardized death rates for lung cancer among men aged 50–69 who were ex-cigarette smokers, by former number of cigarettes smoked per day, and years since last smoking; comparison with death rates for men who were current cigarette smokers, and men who never smoked regularly – American Cancer Society Survey

Ex-cigarette smokers (years since last smoking)	Death rates	
	smoked 1–19 cigarettes/day	smoked 20 + cigarettes/day
Under 1 year	114	283
1–4 years	53	162
5–9 years	20	104
10 + years	7	29
Total ex-smokers	28	101
Current cigarette smokers	120	271
Never smoked regularly	16	16

After HAMMOND [3].

bias is possible: those people with the lowest time exposure to cigarette smoke are the most likely to give up the habit permanently. Further research is needed to resolve this problem.

Cancer Risk with Respect to the Brand of Tobacco

When comparing the studies performed in various countries, the hypothesis appears that the cancer risk would be different according to the kind of tobacco. But firstly we should be sure that there is no bias; however, there may be many other reasons, besides the brand of tobacco, which could explain these differences. Secondly, if there is no bias, we have to try to explain why some tobacco would be more harmful than others. Further studies are needed to try to answer this question.

Relationship between Alcohol and Cancer

The relationship between alcohol and cancer has been particularly studied in France, where – as everybody knows – the consumption of alcohol is very high.

Several geographical pathological surveys have demonstrated a strong correlation between cancer mortality and alcoholism, the latter being meas-

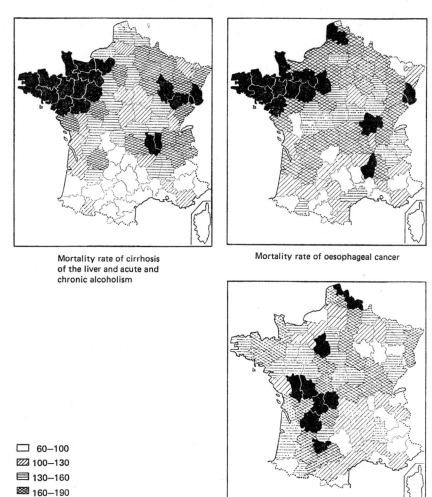

Fig. 2. Comparison of the mortality rate (per 100,000 population, by departments, 1954–1963, men, 45–65 years of age) from alcoholism, oesophageal and rectal cancer. After INSERM.

ured either by cirrhosis of liver mortality or alcohol ingestion [4, 5]. Figure 2 shows clearly that in Brittany and Normandy there is a high mortality both from alcoholism and oesophageal cancer. These mortality data concern the years 1954–1963. It is striking to observe a similar geographical distribution

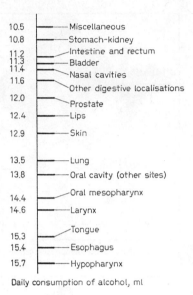

Fig. 3. Classification of cancers according to the level of alcohol consumption. After SCHWARTZ et al. [7].

of oesophageal cancer on mortality data concerning the years 1968–1970, recently published by INSERM [6].

Cancers Associated with Alcohol Consumption

The list of cancers associated with drinking alcohol has been established from several case-control studies. There are mostly oesophagus and hypopharynx and at a lower degree the other sites of the upper respiratory and alimentary tract. Figure 3 shows a classification of the cancers according to the level of alcohol consumption determined from a French survey [7].

Cancer Risk with Respect to the Variety of Alcohol

The study, performed by the IARC in Brittany and Normandy, has specified that the factor probably implied would be the consumption of a home-made alcohol called 'calvados' [8].

But it is necessary to emphasize that no carcinogen factor has ever been found in the alcohols. On the other hand, other clusters of oesophageal cancer have been described, especially in Iran, USSR and China, without the possibility of implicating the alcohol consumption.

Finally, the effect of alcohol appears to be limited to some areas and has an aetiological value less important than tobacco.

Table VI. Sex ratio and the relationship between cancer and the use of tobacco and alcohol in various sites and groups of sites

International classification number	Sites	Number of cases	Sex ratio	Relationship with use of tobacco[1]	Relationship with use of alcohol[1]
147	hypopharynx	4,225	28.0	xx	xx
161	larynx	5,524	27.4	xx	xx
150	esophagus	5,007	16.6	x	xx
162	lung	4,616	11.8	xx	
145	oropharynx	3,216	11.6	xx	x
141	tongue	4,856	9.3	x	xx
143, 144	oral cavity (other sites)	4,145	8.6	xx	x
140	lips	3,609	8.1	x	
181	Bladder and other urinary organs	962	2.6	x	
151	stomach	1,311	2.3		
160	Nasal cavities and accessory sinuses	1,161	2.2		
180	kidney	412	1.8		
155, 157, 158	other digestive sites	274	1.6		
152, 153, 154	intestine and rectum	3,490	1.3		
191	skin	18,561	1.2		
From 192 to 197	nervous system, endocrine glands, various sites	3,465	1.1		
Total		64,834			

[1] This relationship was established as strong (x) and very strong (xx) from previous research. After INSERM [11].

Relationship between both Tobacco, Alcohol and Cancer

It is of course interesting to study simultaneously the two factors, smoking and alcohol consumption, and to distinguish the proper role of each. All the more so because the two habits are closely linked in the general population.

Table VI summarizes the effect of the use both of tobacco and alcohol in a study we carried out on 65,000 cancer cases in France. We can draw the following conclusions: lung and bladder cancer are only related to the use of tobacco; upper respiratory and alimentary cancers are related nearly equally to the use of tobacco and alcohol.

Table VII. Relative risks for oesophageal cancer in male smokers and drinkers in Ille-et-Vilaine, France[1]

		Non- or light smokers (0–9 cig./day)	Moderate smokers (10–19 cig./day)	Heavy smokers (>20 cig./day)	All smokers and non-smokers
Effect of drinking (daily consumption of ethanol in grams)	0–40	1.0	1.0	1.0	1.0
	41–60	5.9	1.8	1.4	3.0
	61–80	9.3	3.2	3.9	5.7
	81–100	12.6	3.2	5.6	6.8
	> 101	22.5	13.2	12.0	16.5
Effect of smoking (daily consumption of ethanol in grams)	0–40	1.0	3.4	5.1	
	41–60	1.0	1.0	1.2	
	61–80	1.0	1.2	2.2	
	81–100	1.0	0.9	2.3	
	> 101	1.0	2.0	2.7	
All levels of alcohol consumption, including non-drinkers		1.0	1.7	2.8	

[1] The upper half of the table, where the effect of increase in daily ethanol intake is examined for each smoking category, should be read vertically. The lower half of the table, where the effect of increase in the amount smoked is examined for each daily ethanol con-category, should be read horizontally. After TUYNS and MASSE [9].

Moreover, it appears that, to a certain extent, the higher the correlation between the sites and the use of tobacco and alcohol, the higher the sex ratio. The conclusion seems to be for us that, at least in countries where the consumption of alcohol is high, the sex ratio reflects the level of the relationship between a cancer site or subsite and the use of tobacco and alcohol. More thorough studies are necessary on this point.

Whenever both factors are implicated in the same cancer, we can investigate the proper role of each. For example, table VII for oesophageal cancer shows that each factor has its own effect, independent of the other [9]. The same results are found for most other upper respiratory and alimentary cancers.

Conclusion

From what has been said, we can ascertain that the tobacco and alcohol cancer group represents almost half of the cancers in the male in France.

It does not seem possible to forsee from the epidemiological data what the decrease in the incidence of cancer would be if the use of tobacco and alcohol were eliminated. We can only note that recently in certain social classes where the use of tobacco had decreased, the rate of lung cancer also decreased.

Summary

Several studies of descriptive and analytic epidemiology have permitted the investigation of cancers linked to the use of tobacco and alcohol.

The tobacco and alcohol cancer group consists of the upper respiratory and alimentary tract, lung and bladder, which accounts for 50% of the cancer in men. When both factors are implicated, the proper role of each could be established.

These data should be considered in decision-making for cancer prevention.

References

1 SCHWARTZ, D.; FLAMANT, R.; LELLOUCH, J., and DENOIX, P.: Results of a French survey on the role of tobacco, particularly inhalation, in different cancer sites. J. natn. Cancer Inst. 26: 1085–1108 (1961).
2 TODD, G.F.: Changes in smoking patterns in the UK. 10th Int. Cancer Congr., Florence 1974.
3 HAMMOND, E.C.: Tobacco; in FRAUMENI Persons at high risk of cancer. An approach to cancer etiology and control, pp. 131–138 (1975).
4 LASSERRE, O.; FLAMANT, R.; LELLOUCH, J. et SCHWARTZ, D.: Alcool et cancer (étude de pathologie géographique portant sur les départements français). Bull. INSERM 22: 53–60 (1967).
5 TUYNS, A.J.: Cancer of the oesophagus – further evidence of the relation to drinking habits in France. Int. J. Cancer 5: 152–156 (1970).
6 Division de la Recherche Médico-Sociale de l'INSERM (Section Cancer et Section Méthodes): Mortalité par cancer en France, 1968, 1969, 1970 (INSERM, Paris 1976).
7 SCHWARTZ, D.; LELLOUCH, J.; FLAMANT, R. et DENOIX, P.: Alcool et cancer. Résultats d'une enquête rétrospective. Revue fr. Etud. clin. biol. 7: 590–604 (1962).

8 Tuyns, A.J. and Masse, G.: Cancer of the oesophagus in Brittany: an incidence study in Ille-et-Vilaine. Int. J. Epidem. *4:* 55–59 (1975).
9 Unit of Epidemiology and Biostatistics: Annual report 1975 (International Agency for Research on Cancer, Lyon 1975).
10 Doll, R.: I-Cancer: Morbidity and mortality attributable to smoking. Int. Congr. on Tobacco and Health, London 1971.

Prof. R. Flamant, Département de Statistique Médicale de l'Institut Gustave Roussy, *Villejuif* (France)

Predictive Carcinogenicity Bioassays in Industrial Oncogenesis

CESARE MALTONI

Institute of Oncology and Tumour Center, Bologna

I. Introduction

It has been estimated that from 80 to 90% of tumours in human beings depend on causes present in the occupational and general environment. Therefore, cancer must be largely considered an ecological disease.

On the other hand, the number and quantity of oncogenic agents in the occupational and general environment have progressively increased in the last few decades, due mainly to the following factors: (1) concentration of oncogenic agents already present in the surface of the Earth; (2) surfacing of oncogenic agents from the depths of the Earth, and (3) production of new potentially oncogenic compounds by chemical and petrochemical industry.

The major problem is now represented by the multitude of products of the synthetic chemical industries, especially of the petrochemical industry. It has been evaluated that, in modern times, from 100,000 to 200,000 new chemical compounds are manufactured annually. Part of these compounds are destined to wide production and diffusion but, since they are new, their effects on animal and human protoplasm are totally unknown.

In this situation, therefore, one should be aware that tools are needed for assessing the risks of the agents that are concentrated, or brought, or *ex novo* introduced into the human environment.

This need is accentuated by the following basic facts: (1) The potentially oncogenic agents in human environment are progressively increasing. (2) Different oncogenic agents may have additive effects. (3) Changes produced by oncogenic agents are largely irreversible. (4) Oncogenic agents may exert their effects on different organs and tissues and widely affect the target organs. (5) A large part of the natural history of tumours takes place without any clinically, and sometimes otherwise detectable, pathologic changes. (6) Cancer is not a reversible disease.

The potential oncogenic risk for man by occupational and environmental agents has been identified, up to recent times, when the incidence of a particular type of tumour in an exposed population has been outstanding, or on the basis of retrospective epidemiological studies.

The gravity of the present situation, however, no longer permits the old policy of 'let us wait and see'.

In other words, the time has come to predict the oncogenic risk in order to avoid the exposure of human beings to it.

The experimental carcinogenicity bioassays (long-term bioassays on laboratory animals), however, are at present far below the need. Up to recent years, they have been hindered by obvious established interests, which incredibly had and still have the ignorant or malicious complicity of part of the scientific community, and which express themselves with: (1) scepticism regarding the validity of the experimental bioassays, based upon past results; (2) fatalistic renounciation because of the huge number of newly produced compounds, and (3) emphasis on the complications involved in elaborate tests.

As regards *scepticism*, we think that the time has come to critically review methods and results of past experiments which, nowadays, may appear impure.

As regards *the large number of compounds*, even if it is true that there are already millions of newly produced chemicals, the ones which urgently need to be tested, however, mainly because of their widespread diffusion, actually number in the hundreds.

Concerning *the complications in performing elaborate tests*, even if it is true that the more experimental tests on animals reproduce the conditions of human exposure, the more relevant they are in man, it should also be pointed out, however, that agents to be examined should follow a pattern of tests which progress in degree of precision and scrutiny, so as to filter out and expose the most dangerous ones.

In sharp contrast to the insufficient attention given to experimental carcinogenicity bioassays for predicting potential risks, there are numerous projects which study the carcinogenic effect on experimental animals and the metabolic patterns of agents which are already known to be carcinogenic for man. These studies, of course, may be of some interest for basic oncology, but they have little bearing on the prevention of occupational and environmental tumours.

II. History

It is a fact that the four most important cases of environmental and occupational tumours, discovered after 1970, have been indirectly or directly predicted in some way, experimentally.

The first is the case of the clear cell adenocarcinomas of the vagina in adolescence, found in girls born from mothers treated during pregnancy with synthetic non-steroid oestrogen therapy. If we go back to 1938, LACASSAGNE had already reported that mammary carcinomas arose in male mice treated with stilbestrol, and in the following years it was shown, by several scientists, that the same hormone was producing a variety of tumours in hormone-dependent and non-dependent tissues, among different experimental animal species [LACASSAGNE, 1950].

The second is the case of pulmonary carcinomas among workers exposed to bis(chloromethyl)ether. The discovery of this new type of occupational tumour came together with experimental evidence: VAN DUUREN et al. [1969] and LASKIN et al. [1971] showed respectively that, when injected subcutaneously into rats, or applied to the skin of mice, the compound was producing subcutaneous fibrosarcomas and skin carcinomas, and when inhaled by rats it induced squamous cell carcinomas of the lung.

The third is the case of lung tumours among workers exposed to chromium pigments.

In March 1973 [MALTONI, 1973], we published the early results of a long range project planned to evaluate on experimental animals the carcinogenic risk of inorganic pigments, among which chromium, molybdenum, cadmium and iron compounds. These results showed that several of the tested compounds, among which lead chromate and lead sulphate, chromate and molybdate, that is chromium orange and yellow and molybdenum orange, were highly carcinogenic.

More than 2 years later, that is in October 1975, the Dry Color Manufacturers' Association made public the data of an epidemiological investigation, started after the experimental evidence, among 580 workers exposed to lead chromate. These data showed that lung cancer was the cause of death for nearly 29% of the deceased workers and accounted for 85% of all cancer deaths.

The fourth case is the history of vinyl chloride carcinogenicity. The carcinogenicity bioassays on vinyl chloride not only had determined the oncogenic potential of this monomer, but they have predicted all the tumours in man which up-to-now have been proved or suspected to be correlated to vinyl chloride exposure (liver angiosarcomas, brain tumours, lung carcinomas, lymphomas and leukaemias and hepatomas), and they have assessed the levels of the risk [MALTONI, 1973; MALTONI and LEFEMINE, 1974a, b, 1975; MALTONI *et al.*, 1974].

III. Approaches and Results

The choice of the agents to be tested should follow a list of priorities based upon many parameters, such as: diffusion of the agent, number of people potentially exposed, direct and/or indirect available data suggesting the risk.

In general, the agents to be tested should be submitted to the easiest and quickest carcinogenicity bioassays, however incomplete those bioassays may be. Then, for those agents inducing tumours in the experimental animals under these conditions, if necessary experiments reproducing the human exposure should be devised, even though they are often sophisticated and costly.

In some cases, because of the importance of the medical and social problems represented by the compound, the need may arise to proceed directly to the most precise, though elaborate, bioassays.

The long-term carcinogenicity bioassays, in a sequence of increasing precision, for extrapolating data to man, may be listed as follows: (1) bioassays for which the route of administration of the agents, and the affected tissues, are not the same as for man (1st level bioassays); (2) bioassays for which the route of administration of the agents is different and the affected tissue is the same as for man (2nd level bioassays); (3) bioassays for which the route of administration of the agents and the affected tissues are the same as for man (3rd level bioassays).

It should be pointed out that the degree of precision is generally directly proportional to the technical complexity.

The potentialities and the possible applications of these three different categories of bioassays will emerge from the following experiments performed in our Institute of Oncology and Tumour Center in Bologna.

1st Level Bioassays

A project of correlated and integrated experiments has been going on for years, to study the oncogenic potentialities of a series of compounds, administered by subcutaneous injection, to rats. The series includes, up to now, chromium, molybdenum, cadmium and iron pigments, other chro-

Table I. Carcinogenicity bioassays of inorganic compounds by subcutaneous injection (30 mg in 1 ml of saline) in Sprague-Dawley rats

Experiment No.	Compound	Number animals at start ♂	♀	total	Progress of the experiment ended	ongoing	weeks from start	Survivors ♂	♀	total	Animals with sarcomas at the site of injection ♂	♀	total number	%
1	Chromite	20	20	40	+		133	0	0	0	0	0	0	–
2	Neochromium (basic chromium sulphate)	20	20	40	+		149	0	0	0	7	3	10	25
3	Chromium alum	20	20	40	+		137	0	0	0	6	2	8	20
4	Chromium yellow (lead chromate)	20	20	40	+		150	0	0	0	10	16	26	65
5	Chromium orange (basic lead chromate)	20	20	40	+		132	0	0	0	14	13	27	67
6	Molybdeneum orange (lead chromate, sulphate and molybdate)	20	20	40	+		117	0	0	0	19	17	36	90
7	Zinc yellow (basic zinc chromate: CrO_3 40%)	20	20	40		+	37	13	5	18	1	0	1	2
8	Zinc yellow (basic zinc chromate: CrO_3 20%)	20	20	40		+	37	17	19	36	1	0	1	2
9	Cadmium yellow (cadmium sulphide)	20	20	40	+		143	0	0	0	9	7	16	40
10	Iron yellow (iron oxide)	20	20	40	+		141	0	0	0	0	0	0	–
11	Iron red (iron oxide)	20	20	40	+		134	0	0	0	1	–	1	2
12	Titanium oxide	60	60	120		+	74	38	47	85	0	0	0	–
13	Magnesium oxide	40	40	80		+	32	32	38	70	0	0	0	–
14	Crocidolite (25 mg)	30	20	50	+		140	0	0	0	1	5	6	12
	Control 1 (to exp. 1–11)	45	15	60	+		124	0	0	0	0	0	0	–
	Control 2 (to exp. 12)	20	20	40		+	74	15	16	31	0	0	0	–
	Control 3 (to exp. 13)	20	20	40		+	32	13	19	32	0	0	0	–
	Control 4 (to exp. 14)	10	10	20	+		139	0	0	0	0	0	0	–

mium compounds, titanium oxide, magnesium oxide and asbestos. The results of several of these experiments, ended or in progress, are given in table I.

These experiments have enabled us to predict the carcinogenic effect of chromium pigments on man, and to make a tentative evaluation of the relative risk of the compounds tested. As far as asbestos is concerned, we have been able to produce further evidence that this agent may be considered a multipotential carcinogen.

Table II. Experiment B08. Treatment: 1 endoperitoneal injection of 25 mg of crocidolite in 1 ml of saline

Groups and treatment	Animals (Sprague-Dawley rats)			Animals with peritoneal mesotheliomas			Animals with metastases	
	sex	number	corrected number[1]	number	%	average latency time (weeks)	number	%
I Crocidolite	♀	50	49	34	69.3	73	13	38.2
	♂	50	48	31	64.5	65	13	41.9
II Saline (control)	♀	45	45	–	–	–	–	–
	♂	15	15	–	–	–	–	–

[1] Animals alive when the first peritoneal tumour arose.

Table III. Experiments BT1, BT6. Exposure by inhalation to VC in air, at 30,000, 10,000, 6,000, 2,500, 500, 250, 50 ppm, 4 h daily, 5 days weekly, for 52 weeks. Results after 135 weeks (end of the experiments)

Groups and treatment	Animals (Sprague-Dawley rats) ♀ and ♂ total	Number of animals with tumours									
		Zymbal gland carcinomas	nephro-blastomas	angiosarcomas		subcutaneous angiomas	skin carcinomas[1]	hepatomas	brain neuro-blastomas	mammary carcinomas	forestomach papillomas
				liver	other sites						
I VC 30,000 ppm	60	35	–	18	1	1	1	1	1	2	11
II VC 10,000 ppm	69	16	5	9	3	4	3	1	7	3	–
III VC 6,000 ppm	72	7	4	13	3	3	1	1	3	–	–
IV VC 2,500 ppm	74	2	6	13	3	3	1	2	5	1	–
V VC 500 ppm	67	4	4	7	2	1	1	3	–	1	–
VI VC 250 ppm	67	–	6	4	2	–	4	–	–	1	–
VII VC 50 ppm	64	–	1	1	1	1	1	–	–	2	–
VIII No treatment	68	–	–	–	–	–	–	–	–	–	–
Total	541	64	26	65	15	13	12	8	16	10	11

[1] Most arising from sebaceous glands.

Table IV. Experiment BT4. Exposure by inhalation to VC in air at 10,000, 6,000, 2,500, 500, 250, 50 ppm, for 4 h daily, 5 days weekly, for 30 weeks. Results after 81 weeks (end of the experiment)

Groups and treatment	Animals (Swiss mice) ♀ and ♂ total	sur-vivors	Number of animals with tumours					
			pulmonary tumours	mammary carcino-mas[1]	liver angio-sarcomas	vascular tumours of other type and/or site	epithelial tumours of the skin[2]	fore-stomach papillo-mas
I VC 10,000 ppm	60	–	35	13	8	9	3	1
II VC 6,000 ppm	60	–	38	8	5	9	6	1
III VC 2,500 ppm	60	–	30	9	11	12	3	1
IV VC 500 ppm	60	–	38	7	11	18	1	–
V VC 250 ppm	60	–	33	11	11	22	2	–
VI VC 50 ppm	60	–	2	12	1	13	–	–
VII No treatment	150	–	8	–	–	1	–	–
Total	510	–	184	60	47	84	15	3

[1] Most of which with squamous metaplasia.
[2] Some arising from sebaceous glands.

2nd Level Bioassays

An experiment was performed to assess the oncogenic potential of asbestos on peritoneal mesothelium, by injecting crocidolite directly into the abdominal cavity of rats. The results shown in table II are consistent with the well-known epidemiological results which point to the risk of developing peritoneal mesotheliomas among asbestos workers, which lately has been becoming more and more prominent.

3rd Level Bioassays

17 different experiments have been performed to study the carcinogenic potential of vinyl chloride (VC), administered through different routes (in-

Table V. Experiment BT8. Exposure by inhalation to VC in air at 10,000, 6,000, 2,500, 500, 250, 50 ppm, 4 h daily, 5 days weekly, for 30 weeks. Results after 109 weeks (end of the experiment)

Groups and treatment	Animals (golden hamsters) total	Number of animals with tumours						
		liver angio-sarcomas	skin trichoepitheliomas and basaliomas[1]	melanomas	lymphomas	fore-stomach epithelial tumours[2]	other type and/or site	total[3]
I VC 10,000 ppm	35	–	6	1	–	4	2[4]	10
II VC 6,000 ppm	32	1	2	2	2	7	7[5]	10
III VC 2,500 ppm	33	–	1	1	1	11	3[6]	13
IV VC 500 ppm	33	2	4	–	1	7	2[7]	12
V VC 250 ppm	32	–	3	–	1	2	–	6
VI VC 50 ppm	33	–	6	1	1	4	–	10
VII No treatment	70	–	2	–	2	2	–	7
Total	268	3	24	5	8	37	14	68

[1] Several cases with acanthosis and some undergoing malignant transformation.
[2] Papillomas, acanthomas, some of which undergoing malignant transformation.
[3] Several animals with 2 or more tumours.
[4] 1 subcutaneous angioma; 1 gall-bladder adenocarcinoma.
[5] 2 hepatomas; 2 liver fibroangiomas; 2 liver angiomas; 1 biliducts adenocarcinoma.
[6] 1 hepatoma; 1 liver fibroangioma; 1 liver angioma.
[7] 1 subcutaneous angioma; 1 bronchial carcinoma.

halation, ingestion and peritoneal and subcutaneous injection), at different concentrations, for varying periods of time, by continuous or intermittent treatment, on animals of different species (rats, mice, hamsters), strains (Sprague-Dawley and Wistar rats), sex and age (adults, new-borns, embryos).

When given by inhalation (the most important route of human exposure), the monomer causes, in one or more of the tested species, the following tumours: angiosarcomas and angiomas of the liver and other sites, nephro-

Table VI. Experiment BT15. Exposure by inhalation to VC in air, at 25, 10, 5, 1 ppm, 4 h daily, 5 days weekly, for 52 weeks. Results after 87 weeks

Groups and treatment	Animals (Sprague-Dawley rats)		Number of animals with tumours						
	total	survivors	Zymbal gland carcinomas	nephroblastomas	angiosarcomas liver	angiosarcomas other sites	mammary carcinomas	other type and/or site	total
I VC 25 ppm	120	45	3	–	3	–	10	7	20
II VC 10 ppm	120	51	1	–	–	–	11	5	16
III VC 5 ppm	120	62	–	–	–	–	13	5	17
IV VC 1 ppm	120	49	–	–	–	–	8	4	12
V No treatment	120	49	–	–	–	–	2	4	6
Total	600	256	4	–	3	–	44	25	71

blastomas, brain neuroblastomas, Zymbal gland carcinomas, skin tumours, mammary carcinomas, forestomach papillomas, lung adenomas, hepatomas and others (tables III–V).

Since April 1974, VC appears to be effective up to 50 ppm. We know now that VC, under our experimental conditions, is carcinogenic in rats also at lower doses, namely at 25 ppm (table VI).

As already mentioned, experimental bioassays, in the case of VC, have predicted the carcinogenicity of the compound and given indication of the target organs. Moreover, they have supplied information on the levels of risk, in relation to the dose, which provides the basis for assessing the MAC standard level, the MAC time period and the types of monitoring.

IV. Conclusions

We wish, as we did in the past, that proper experimental bioassays would be performed following a list of priorities on all the agents already produced, used and widespread in the human environment, and on the new ones destined for production.

Projects of carcinogenicity bioassays on several other compounds produced and diffused on a large scale are now going on in our Institution, including: (1) monomers used in plastic industries, such as styrene, acrylonitrile, vinylidene chloride and others; (2) polymers; (3) propellents; (4) oestrogen and progestogen hormones, and (5) petrol proteins.

References

DUUREN, B.L. VAN; SIVAK, A.; GOLDSCHMIDT, B.M.; KATZ, C., and MELCHIONNE, S.: Carcinogenicity of halo-ethers. J. natn. Cancer Inst. *43:* 481–486 (1969).

LACASSAGNE, A.: Apparition d'adénocarcinomes mammaires chez des souris mâles traitées par une substance oestrogène synthétique. C.r. Séanc. Soc. Biol. *129:* 641 (1938).

LACASSAGNE, A.: Les cancers produit par des substances chimiques endogènes (Herman, Paris 1950).

LASKIN, S.; KUSCHNER, M.; DREW, R.T.; CAPPIELLO, V.P., and NELSON, N.: Tumors of the respiratory tract induced by inhalation of *bis*(chloromethyl)ether. Archs envir. Hlth *23:* 135–136 (1971).

MALTONI, C.: Occupational carcinogenesis. 2nd Int. Symp. on Cancer Detection and Prevention, Bologna 1973; in MALTONI Advances in tumour prevention, detection and characterization, vol. 2, pp. 19–26 (Excerpta Medica, Amsterdam 1974).

MALTONI, C. and LEFEMINE, G.: Carcinogenicity bioassays of vinyl chloride. I. Research plan and early results. Envir. Res. *7:* 387–405 (1974a).

MALTONI, C. e LEFEMINE, G.: Le potenzialità dei saggi sperimentali nella predizione dei rischi oncogeni ambientali. Un esempio: il cloruro di vinile. Accademia Nazionale dei Lincei, Rendiconti della Classe di Scienze Fisiche, Matematiche e Naturali, vol. LVI, serie VIII, fasc. 3 (1974b).

MALTONI, C. and LEFEMINE, G.: Carcinogenicity bioassays of vinyl chloride: current results; in SELIKOFF and HAMMOND Toxicity of vinyl chloride-polyvinyl chloride, pp. 195–218 (New York Academy of Sciences, New York 1975).

MALTONI, C.; LEFEMINE, G.; CHIECO, P. e CARRETTI, D.: La cancerogenesi ambientale e professionale: nuove prospettive alla luce della cancerogenesi da cloruro di vinile. Osped. Vita *1:* 5–6, 4–66 (1974).

Dr. C. MALTONI, Institute of Oncology and Tumour Center, *Bologna* (Italy)

Measurement of Chemical Carcinogens in the Human Environment: the Objectives and Problems Encountered

LAIMA GRICIUTE

Unit of Environmental Carcinogens, IARC, Lyon

Introduction

Considerable data on the carcinogenicity of chemical substances have been obtained in the laboratory. There are several hundred chemicals which have proved to be carcinogenic in animals. Some of them are widely distributed in the human environment, but we do not know whether they represent a danger to man, since the data obtained from animal experiments cannot be directly extrapolated to human pathology. Some chemical substances certainly play an important role in the development of human cancers; according to the IARC monograph series[1], about 20 have so far proved to be carcinogenic in man.

The variations in cancer incidence between different regions of the world give the impression that the majority of human cancers (around 80% according to HIGGINSON [1969]), are due to environmental factors. The relative lack of importance of genetic differences between races and populations can be deduced from data on cancer incidence in migrants [KMET, 1970], as the cancer morbidity of these populations changes over time and becomes similar to that of the country to which they migrate. In spite of the suggestion that some of these geographical differences are caused by carcinogenic agents in the environment, there is still a considerable gap between the data obtained in carcinogenesis laboratories and its relation to human cancer morbidity.

Experimental results indicate that in carcinogenesis the amount of chemical plays an essential role; the numbers of induced tumours, the number of tumour-bearing animals, the period of latency and even the degree of malig-

[1] *IARC Monographs on the Evaluation of Carcinogenic Risk of Chemicals to Man*, vol. 1–12.

nancy depend on the amount of carcinogen used. Dose-response relationships are clear when assaying chemicals of any group: benzo[α]pyrene [JANYSHEVA, 1970], N-nitrosamines [TERRACINI et al., 1967], ethyl urethane [GRICIUTE, 1962], DDT [TURUSOV et al., 1973] and others.

One of the reasons why direct extrapolation of animal data to human pathology is impossible is the lack of comparability of levels of dosage. The quantities given in animal experiments represent some sort of overloading, out of proportion to the quantities to which man is exposed.

The information available on the quantities of carcinogens identified in the human environment, apart from certain industrial situations, are quite incomplete. Scattered information on the contamination of air, water, soil and food by polycyclic aromatic hydrocarbons (PAH) as well as some data on volatile N-nitrosamines and their precursors or mycotoxins in food have been obtained. These measures have been acquired using a variety of methods for sampling, storage, clean-up and identification; they are therefore hardly comparable and cannot give an exact idea of the exposure of man to these widely distributed chemical carcinogens. The problem is further complicated by the fact that very little is known on the background levels of chemical carcinogens in the human environment.

The measurement of environmental chemical carcinogens is important for the following reasons: (1) the data acquired would be valuable for the evaluation of the role of some carcinogens in human cancer; (2) it would provide some indication of the present-day levels of widely distributed chemical carcinogens against which to gauge future control measures; (3) it would provide a scientific and technical basis for reduction of environmental chemical carcinogens in both the near and distant future.

The point of departure for the evaluation of the role of chemical carcinogens in human cancer must be epidemiological data on cancer incidence. Demonstration of a correlation between cancer morbidity patterns and distribution of known chemical carcinogens and *per capita* intake should lead to the effective control of these substances in the environment.

Priorities

1. Substances

Priorities must be established for this approach to the study of chemical carcinogens.

It is more feasible to establish correlations for industrial chemical carcinogens since the concentration is as a rule higher, the chemical and source more readily identified, for many background levels usually do not exist. But the number of cases of industrial cancer compared to all cancers is not high: 1–5% [CLAYSON, 1970], 12% [HIRAYAMA, 1975], 20% [BOYLAND, personal commun., 1975]. Thus, although study and control is more feasible in the industrial setting, it does not solve the overall question of general population exposure to environmental carcinogens. Widely distributed carcinogens (polycyclic aromatic hydrocarbons, N-nitroso compounds, mycotoxins) should therefore receive our greatest attention.

The only realistic approach would be the analysis of environmental substrates for known chemical carcinogens and the estimation of the total load. But sometimes the prevalence of one carcinogen, which obscures the action of others, can be assumed [TOMATIS, 1969].

2. Areas

The determination of priorities for the areas where estimations are to be undertaken is a very important question [MUIR, 1975a, b].

When measuring the exposure of a population in the developed countries it is likely that similar levels of widely distributed carcinogens will be found, as well as similar levels of cancer incidence. Such data would be of little value to establish a causal relationship [HIGGINSON, 1972].

The regions most suited to this type of study would be those where traditional ways of life have been maintained and where peculiar patterns of cancer morbidity have been found. The populations in such regions represent one kind of high-risk group.

Analytical studies carried out in the areas of the world where cancer morbidity patterns are unusual can indicate which substances could be implicated in the development of some particular cancer as a causal factor. On the other hand, data from analytical studies, showing no correlation with morbidity rates, would indicate that the levels of chemical carcinogens obtained were probably not dangerous.

One of the difficulties encountered when measuring chemical carcinogens is the fact that the present-day environment may no longer resemble the environment which some 15 years ago, for example, triggered off the cancer mechanism. The fact that the environment is constantly changing should not obviate the usefulness of the measurement of chemical carcinogens in the environment, for the following reasons.

First, it is doubtful whether the most important substrates in the environment are changing very rapidly. It is certain that every year new chemicals join the list of pollutants, but the disappearance with no special effort being made of substances such as polycyclic aromatic hydrocarbons or *N*-nitrosamines is doubtful.

Secondly, the measurement of chemical carcinogens in the environment could be used as baseline data in prospective epidemiological studies.

Problems

The problems which arise when measuring known chemical carcinogens fall into two categories: (1) inadequate techniques; (2) difficulties in interpretation of data acquired.

1. Measurement

The measurement of chemical substances presents considerable difficulties for the following reasons: (1) the levels of chemical carcinogens in the environment are generally low. The methods of estimation must be very sensitive. (2) These chemicals can contaminate every substrate in the environment: air, food, water. Therefore, the sampling, separation from the substrate, and clean-up require different procedures. (3) Each environmental substrate can be contaminated by different substances from the same and different chemical groups. For estimation of the total load of chemical carcinogens, the substrate should be analysed for different chemicals. In some groups of known carcinogens, representative compound-indicators have been chosen, e.g. benzo(α)pyrene is considered as an indicator of the group of polycyclic aromatic hydrocarbons. For other groups, there is no such indicator.

Considerable variations exist in the levels of accuracy of the analytical methods available. One of the most important steps, therefore, in the establishment of a measuring system for chemical carcinogens in the environment is the design of sensitive and reproducible analytical methods. Analyses of environmental substrates will only be of value if standardized methods of sampling and analysis are applied, to permit comparison of results obtained by different laboratories throughout the world.

Some IARC Studies

Nitrosamines. In order to contribute to the standardization of the methods for the analyses of volatile *N*-nitrosamines, IARC has started a cooperative study, 3 stages of which have just been completed.

Table I. Statistical parameters for analytical results on samples of canned meat spiked at the 20-μg/kg level

	NDMA	NDEA	NDBA	NPy
True value	6.3	5.2	6.6	6.6
Overall mean	5.07	4.61	4.84	4.5
Average within lab. SD	0.954	0.750	0.919	1.98
Corresponding coefficient of variation, %	18.8	16.2	18.9	44
SD of means obtained by different labs.	1.98	1.8	2.57	2.68
Corresponding coefficient of variation, %	39	39	53.1	59.5

Stage I: Aqueous and methylene chloride solutions of 10 μg/kg N-nitrosodiethylamine and nitrosopyrrolidine were sent to the laboratories willing to collaborate. Everyone chose his own method of analysis. The results were evaluated in IARC and were encouraging.

Stage II: Luncheon meat spiked with N-nitrosodimethylamine, N-nitrosodiethylamine, N-nitrosodibutylamine and N-nitrosopyrrolidine at the 20 μg/kg level were distributed.

The results from 15 laboratories have shown an interlaboratory reproducibility of 50% and an intralaboratory reproducibility of 20% [WALKER and CASTEGNARO, 1974]. These data are considered as satisfactory, since the carcinogens are present in the analysed substances in almost trace levels.

Stage III: The substance analysed was the same, but the levels of N-nitrosamines were lower – 5 μg/kg, which is nearer to environmental levels usually encountered. Reproducibility was almost the same (table I). Today, everyone agrees that the most suitable method for identification of volatile N-nitrosamines is gas chromatography with mass spectrometry confirmation [WALKER and CASTEGNARO, 1976].

In spite of technical difficulties, some field studies have been carried out. One of them was the measurement of aflatoxin levels. It was assumed that very potent liver carcinogens, aflatoxins, could be associated to endemic liver cancer in Africa. Furthermore, it was established that aflatoxin is commonly present in local ground-nuts, which is the staple food in these African countries where the morbidity of liver cancer is unusually high (30:100,000).

The analysis of foods for aflatoxin in Kenya, Swaziland and Mozambique has been undertaken. The results of all these studies indicated the

Table II. Daily aflatoxin intake and liver cancer incidence in Kenya[1]

Parameter	Altitude sub-area		
	high	middle	low
Frequency of contaminated diets	39/808	54/808	78/816
Mean aflatoxin level, µg/kg	0.121	0.205	0.351
Frequency of contaminated beer	3/101	4/101	9/102
Mean aflatoxin level, µg/l	0.050	0.069	0.167
Mean aflatoxin ingested[2], ng/kg body weight per day	4.88	7.84	14.81
Liver cancer incidence per 100,000 per year	3.11	10.80	12.12

[1] Data are for males aged 16 or over. Source: International Agency for Research on Cancer, 1972.
[2] Calculated on the assumption of a 2-kg intake of food and a 2-litre intake of beer per day, and an average body weight of 70 kg.

Table III. Quantities of dimethylnitrosamine in 179 samples of Iranian food

	High incidence, % (88 samples)		Low incidence, % (91 samples)	
Not detected	25	(22)	35.3	(32)
Detected	75	(66)	64.8	(59)
1–5 µg/kg	59	(52)	51.6	(47)
<1 µg/kg	16	(14)	13.1	(12)

Table IV. Summary of levels of benzo[α]pyrene in bread (high incidence area) and cooked rice (low incidence area)

	Benzo[α]pyrene, µg/kg		
	average	maximum	minimum
Bread	0.4	1.1	< 0.2
Cooked rice	0.13	0.7	< 0.2

Minimum level of detection: 0.2 µg/kg.

positive correlation between the level of aflatoxin intake and liver cancer morbidity (table II) [LINSELL and PEERS, 1972; VAN RENSBURG et al., 1974; PEERS et al., 1976]. Similar studies in Thailand show the same correlation [WOGAN, 1973]. Therefore, even if we cannot consider aflatoxin intake as the sole cause, it is quite certain that aflatoxins play an important role in endemic liver cancer development.

The data are simpler to evaluate when the population group is mostly exposed to a single identified carcinogen. But such a situation seldom occurs. People are usually exposed to different chemical carcinogens at the same time.

Results which are not easy to interpret have been obtained when analysing traditional food sampled in Iran on the Caspian littoral in areas of different morbidity of oesophageal cancer.

The incidence of oesophageal cancer is specially high in the eastern part of the Caspian littoral (109/100,000 for men and 174/100,000 for women) inhabited by Turkomans, and relatively low in the western part (19/100,000 for men, 7/100,000 for women) inhabited by non-Turkomans [KMET and MAHBOUBI, 1972]. The traditional food was sampled in 4 villages of both areas, and 179 samples were analysed for chemical carcinogens. Special attention was paid to volatile N-nitrosamines since they are organ-specific for the oesophagus in animal experiments. The staple food (50 samples), which is different in the two areas – bread being the staple food in the high-morbidity area and rice in the low-morbidity area – was analysed for PAH. Analyses of food samples for aflatoxins was also performed.

Some N-nitrosamines, mostly N-dimethylnitrosamine (which is a liver carcinogen in animal experiments) were detected. But the level of these substances was rather low and differences between the two areas were not significant (table III). PAH was also detected in bread in small quantities (table IV). There were less PAH in rice. Aflatoxin was present in only one sample [IARC Annual Report, 1974, 1975].

In other studies on high rates of oesophageal cancer morbidity in the north of France, a high consumption of local home-brewed apple brandy was implicated [TUYNS, 1975]. Analyses for known carcinogens, namely volatile N-nitrosamines and PAH, are currently being carried out. The results obtained so far seem interesting, because N-nitrosamines and benzo(α)pyrene have been detected in home-brewed brandy, and not in apple brandy of industrial origin (tables V and VI) [IARC Annual Report, 1975, 1976].

Further analyses are necessary before it can be stated with certainty that chemical carcinogens present in alcoholic drinks are responsible for the development of the cancers related to alcohol consumption.

Table V. Results of analysis for nitrosamines of apple brandies and other alcoholic beverages

Sample	Origin of sample	NDMA, µg/kg
Farm apple brandies	Finistère (Brittany)	2.5
Farm apple brandies	Finistère (Brittany)	2
Farm apple brandies	Finistère (Brittany)	3
Farm apple brandies	Finistère (Brittany)	ND[1]
Farm apple brandies	Finistère (Brittany)	1
Farm apple brandies	Finistère (Brittany)	ND
Farm apple brandies	Finistère (Brittany)	1
Farm apple brandies	Finistère (Brittany)	ND
Farm apple brandies	Finistère (Brittany)	10*
Farm apple brandies	Finistère (Brittany)	ND
Farm apple brandies	Finistère (Brittany)	0.5
Farm apple brandies	Finistère (Brittany)	2
Farm apple brandies	Finistère (Brittany)	2
Farm apple brandies	Finistère (Brittany)	ND
Farm apple brandies	Finistère (Brittany)	ND
Farm apple brandies	Finistère (Brittany)	ND
Farm apple brandies	Finistère (Brittany)	ND
Farm apple brandies	Finistère (Brittany)	ND
Farm apple brandies	Finistère (Brittany)	ND
Farm apple brandies	Rennes (Brittany)	ND
Farm apple brandies	Rennes (Brittany)	3.5
Farm apple brandies	Rennes (Brittany)	1
Farm apple brandies	Rennes (Brittany)	5
Farm apple brandies	Rennes (Brittany)	2
Farm apple brandies	Rennes (Brittany)	ND
Farm apple brandies	Rennes (Brittany)	1
Farm apple brandies	Rennes (Brittany)	ND
Farm apple brandies	Rennes (Brittany)	ND
Farm apple brandies	Quimper-Concarneau (Brittany)	2
Farm apple brandies	Quimper-Concarneau (Brittany)	ND
Commercial calvados	Quimper-Concarneau (Brittany)	ND
Commercial calvados	Quimper-Concarneau (Brittany)	ND
Apple brandy	USA (one manufacturer)	ND
Apple brandy	USA (one manufacturer)	ND
Apple brandy	USA (one manufacturer)	ND
Apple brandy	USA (one manufacturer)	ND
Apple brandy	USA (one manufacturer)	ND
Apple brandy	USA (one manufacturer)	ND
Apple brandy	USA (one manufacturer)	ND
Apple brandy	USA (one manufacturer)	ND
Apple brandy	USA (one manufacturer)	ND
Apple brandy	USA (one manufacturer)	ND

[1] ND: not detected.

Table VI. Results for determination of PAH

Sample	Benzo(α)pyrene, µg/kg
Calvados of various origins (Brittany)	5–10
	1–5
	5–10
	5–10
	5–10
	5–10
	5–10
	5–10
	5–10
	5–10
	5–10
	5–10
	5–10
	5–10
	5–10
Apple brandy (USA one manufacturer)	ND
	ND
	ND
	ND
	ND
	ND
	ND
	ND
	ND
	ND

2. Interpretation

The interpretation of measurements is not feasible at present for widely distributed chemical carcinogens for the following reasons: (1) the 'effective' and 'no-effect' doses of chemical carcinogens for man have not been established; (2) the 'reference' levels for widely distributed carcinogens (PAH, nitrosamines, mycotoxins, etc.) have not yet been determined, and hence the significance of other or additional exposures is difficult to evaluate; (3) the mode of action of small doses of combined carcinogens, which would represent the real situation in the human environment, is unknown, even in experimental animals.

The first two problems can be solved only by gathering the quantitative data on chemical carcinogens in different environments, establishing correlations with morbidity and comparing the differences. The third problem can be approached by experimental studies on the chronic intake of small doses of different chemical carcinogens.

Discussion

The ultimate goal of quantitative estimation of chemical carcinogens in the environment is the prevention of cancer. The tool which should lead to this goal is the monitoring of chemical carcinogens in the human environment. This is not currently possible, but the efforts of numerous cancerologists and environmental health scientists are directed towards an approach to monitoring.

One of the requirements for the establishment of control and limitations of any substances is the acceptance of the concept that there is a threshold of harmful action.

Some cancerologists are of the opinion that because there is no mechanism for determining the existence of biological thresholds for chemical carcinogens, the concept of a threshold limit value is totally inapplicable to carcinogenesis [EPSTEIN, 1975].

Others [LATARJET, 1976] consider that even if theoretically in carcinogenesis as well as in mutagenesis, a threshold level does not exist, there is some practical threshold. Development of cancer in man is not only due to cell transformation. Therefore, the limitation of chemical carcinogens which for one reason or another cannot be withdrawn from the human environment is very important.

TOMATIS [1969] wrote that our goal can only be the reduction of environmental carcinogens to unavoidable background levels.

The best example of a fruitful multidisciplinary approach to the study of one group of chemical carcinogens, which subsequently led to some control and legislative measures, is for PAH – the first group of chemical carcinogens to be studied biologically and found to be responsible for certain occupational cancers.

The case reports attributable to mixtures containing PAH, and numerous experimental data on the carcinogenicity of pure PAH has confirmed the importance of PAH in human pathology. The stability of these compounds, and especially of benzo(α)pyrene, and some of their physical properties which have enabled their identification without difficulty, have encouraged the creation of monitoring systems for their detection in the environment.

Benzopyrene, which is considered as an indicator for all PAH groups, is one of the carcinogens about whose distribution and levels in the environment we know the most. It has been established that the background level for benzopyrene in the soil is 5–6 μg/kg [SHABAD, 1971].

PAH is at present monitored in the air of towns and industrial regions of developed countries [SHABAD and DIKUN, 1959; LAWTHER and WALLER, 1975]; it is not measured in other environmental substrates to such an extent and this will have to be done in the future.

It is not surprising, therefore, that PAH, and in particular benzopyrene, was the first group of widely distributed carcinogens for which a 'maximum allowable concentration' was established: 0.1 μg/100 m^3 in the ambient air; 15 μg/100 m^3 in the air of work places [SHABAD, 1971, 1975], and 0.2 μg/litre of 6 PAH in water [WHO, 1970].

It is important to realize that the 'maximum allowable concentration' can only be considered as a working hypothesis at this stage of our knowledge and the studies must be continued. We do not have a sound scientific basis for legislation, but the opinion that nothing can be done is a rather defeatist attitude.

The epidemiological data on cancer morbidity are vital in the studies on environmental carcinogenesis. Epidemiology can offer the working hypothesis, can ask the questions, but it cannot answer them without help from other disciplines. We hope that the analytical laboratory can contribute largely in answering the questions asked by the epidemiologists.

Acknowledgements

I wish to thank Dr. A.C. LINSELL, Dr. C.S. MUIR and M.M. CASTEGNARO for helpful criticism, and Mme M. COURCIER for typing the manuscript.

References

CLAYSON, D.B.: Chemicals as carcinogens in man. Yorks. Faculty J. *10:* 1–5 (1970).
EPSTEIN, S.: Regulatory aspects of occupational carcinogens: contrast with environmental carcinogens. Proc. IARC-INSERM Symp. on Environmental Pollution and Carcinogenic Risks, Lyon 1975.
GRICIUTE, L.: On the mechanism of the action of urethane on lung tissue (in Russian). Pathol. Physiol. exp. Ther. *2:* 69–70 (1962).

HIGGINSON, J.: Present trends in cancer epidemiology; in MORGAN Can. Cancer Conf., pp. 40–75 (Pergamon Press, Oxford 1969).

HIGGINSON, J.: The role of geographical pathology in environmental carcinogenesis. Environment and cancer, pp. 70–92 (Williams & Wilkins, Baltimore 1972).

HIRAYAMA, T.: A sound evaluation of the problem of occupational and industrial cancer in the present-day society. Working paper for the Scientific Group on Methods of Monitoring Carcinogenic Chemicals in the Environment, Geneva 1975.

JANYSHEVA, N.: Sanitary protection of environment from chemical carcinogens contaminating the wastes of industry (in Russian), doct. thesis, Moscow (1970).

KMET, J.; cited by MUIR Pointers from regional variations in incidence – international perspectives; in GRUNDMANN and PEDERSEN Recent results in cancer research, vol. 50, pp. 132–140 (Springer, Berlin 1975).

KMET, J. and MAHBOUBI, E.: Oesophageal cancer in the Caspian littoral of Iran: initial studies. Science *172:* 846–853 (1972).

LATARJET, R.: Réflexions sur la pollution cancérigène par les radiations ionisantes. Bull. Cancer *63:* 1–10 (1976).

LAWTHER, P.J. and WALLER, R.E.: Coal fibers, industrial emissions and motor vehicles as sources of environmental carcinogens. Proc. IARC-INSERM Symp. on Environmental Pollution and Carcinogenic Risks, Lyon 1975.

LINSELL, C.A. and PEERS, F.G.: The aflatoxins and human liver cancer; in GRUNDMANN and TULINIUS Current problems in the epidemiology of cancer and lymphomas, pp. 125–129 (Springer, Berlin 1972).

MUIR, C.S.: Pointers from regional variations in incidence – international perspectives; in GRUNDMANN and PEDERSEN Recent results in cancer research, vol. 50, pp. 132–140 (Springer, Berlin 1975a).

MUIR, C.S.: Possibility of monitoring populations to detect environmental carcinogens. Proc. at IARC-INSERM Symp. on Environmental Pollution and Carcinogenic Risks, Lyon 1975b.

PEERS, F.G.; GILMAN, G.A., and LINSELL, C.A.: Dietary aflatoxins and human liver cancer – a study in Swaziland. Int. J. Cancer *17:* 167–176 (1976).

RENSBURG, S.J. VAN; WATT, J.J. VAN DER; PURCHASE, J.F.H.; PEREIRA CONTINHO, L., and MURKHAM, R.: Primary liver cancer rate and aflatoxin intake in a high cancer area. S. Afr. med. J. *48:* 2508a–2508d (1974).

SHABAD, L.: On the possibility of hygienic limitation of carcinogenic substances – 'maximum allowable concentration' evaluation (in Russian). Gig. Sanit. *10:* 93 (1971).

SHABAD, L.: On the so-called MAC for carcinogenic hydrocarbons. Neoplasma *22:* 459–468 (1975).

SHABAD, L. and DIKUN, P.P.: Atmospheric air pollution with benzo[a]pyrene (in Russian). Medzig, Leningrad (1959).

TERRACINI, B.; MAGEE, P.N., and BARNES, J.M.: Hepatic pathology in rats on low dietary levels of dimethylnitrosamine. Br. J. Cancer *21:* 559–565 (1967).

TOMATIS, L.: The exposure of humans to the total environmental carcinogenic load. Proc. 15th Int. Congr. Occupational Health, Tokyo, 1969, pp. 268–270.

TURUSOV, V.S.; DAY, N.E.; TOMATIS, L.; GATI, E., and CHARLES, R.T.: Tumors in CF-1 mice exposed for six consecutive generations to DDT. J. natn. Cancer Inst. *51:* 983–997 (1973).

TUYNS, A.J. et MASSE, G.: Le cancer de l'oesophage en Ille-et-Vilaine. Ouest méd. 5: 1757–1770 (1975).

WALKER, E.A. and CASTEGNARO, M.: A report on the present status of a collaborative study of methods for the trace analysis of volatile nitrosamines; in BOGOVSKI and WALKER N-Nitroso compounds in the environment. IARC Scientific Publications No. 9, pp. 57–62 (International Agency for Research on Cancer, Lyon 1974).

WALKER, E.A. and CASTEGNARO, M.: New data on collaborative studies on volatile nitrosamines; in WALKER, BOGOVSKI and GRICIUTE Environmental N-nitroso compounds, analysis and formation. IARC Scientific Publications No. 14 (International Agency for Research on Cancer, Lyon 1976).

WOGAN, G.N.: Assessment of exposure to aflatoxins; in DOLL and VODOPIJA Host environment interactions in the etiology of cancer in man, IARC Scientific Publications No. 7, pp. 237–242 (International Agency for Research on Cancer, Lyon 1973).

WHO: European standards for drinking water; 2nd ed. (WHO, Genève 1970).

Dr. L. GRICIUTE, Chief, Unit of Environmental Carcinogens, IARC, 150, Cours Albert Thomas, *F-69008 Lyon* (France)

Aging, Carcinogenesis and Radiation Biology

KENDRIC C. SMITH

Department of Radiology, Stanford University School of Medicine, Stanford, Calif.

From radiation biology, we have learned that deoxyribonucleic acid (DNA) is the most important single molecule within a cell [1, 2]. If DNA is damaged and not repaired properly, then mutations and/or death result. All other molecules within a cell are less important because they are present in multiple copies, or can be replaced if the genetic information residing in the DNA remains intact.

From radiation chemistry and photochemistry, we have learned the types of chemical changes that can occur in DNA [3, 4]. These include chain breaks, and damage to the purine and pyrimidine bases. Chain breaks can occur at seven chemically different sites along the DNA chain, and undoubtedly call into play different groups of enzymes for their repair.

Base damage can be divided into two categories, unimolecular and bimolecular. There is a finite number of unimolecular reactions that can occur in the purines and pyrimidines, but there is almost an infinite number of bimolecular reactions, since these depend upon the particular environment of the DNA. Bimolecular addition reactions with DNA can involve normal cellular components such as other nucleic acid bases, proteins, amino acids, and other metabolites, and they can involve exogenous compounds such as drugs [5]. The chemistry and biological consequences of these DNA addition reactions was the subject of a recent international symposium [6].

From radiation biology, we have learned of the capacity of cells to repair or circumvent these chemical changes in DNA [7]. Two major categories of DNA repair are known: excision repair and post-replicational repair. In both of these systems, which are for the repair of DNA base damage, enzymatically induced DNA single-strand breaks are produced that are subsequently repaired. Similar repair steps are also involved in the repair of DNA single-strand breaks that are produced directly by ionizing radiation [8]. In *Escherichia coli*, DNA double-strand breaks appear to be unrepairable and lethal [9]. There still remains a controversy as to whether DNA double-strand breaks are repaired in mammalian cells or not [10 and references therein].

In excision repair, the damaged DNA bases are cut and replaced with undamaged nucleotides. In bacteria there are two pathways for excision repair, a major one that seems to be largely error free, and a minor one that appears to be error prone, and leads to the production of mutations [11].

In post-replication repair, the damage is bypassed during DNA replication leaving gaps in the newly replicated daughter-strand DNA. These gaps are then filled, at least in part, by material cut from the parental strands of DNA. In bacteria, there exist at least five separate enzymatic pathways of post-replication repair [12].

Post-replication repair appears to be more mutagenic than is excision repair. This is based upon the observation that cells deficient in excision repair are more easily mutagenized by ultraviolet (UV) radiation than are wild-type cells [13]. This observation has now been confirmed with mammalian cells [14].

The role of DNA repair in the production of mutations is most dramatically exemplified by the observation that bacterial cells that are genetically deficient in post-replication repair, as well as the minor pathway of excision repair (i.e. *recA* and *lexA* [*exrA*] strains), cannot be mutagenized by UV radiation [13]. Thus, it seems clear that mutagenesis must be due primarily to errors in the repair of damaged DNA.

Presumably, chemical carcinogenesis is also the result of the error-prone repair of chemically damaged DNA. This concept is supported by the observation that nearly all chemical carcinogens are mutagens [15]. The mutagenic theory of carcinogenesis is receiving ever-increasing support [16].

Evidence to support the role of DNA repair in carcinogenesis is based, in part, upon the observations that patients with heritable syndromes that greatly increase their chances of cancer have deficiencies in DNA repair. Thus, patients with the syndrome xeroderma pigmentosum show a high incidence of cancer, and are deficient in certain types of repair of damage produced by UV radiation [17], while patients with ataxia telangiectasia are deficient in the repair of certain types of DNA damage produced by ionizing radiation [18]. Presumably these deficiencies in repair are of the error-free type leaving the lesions to be repaired by error-prone, mutagenic pathways of repair. In confirmation of this hypothesis, xeroderma pigmentosum cells

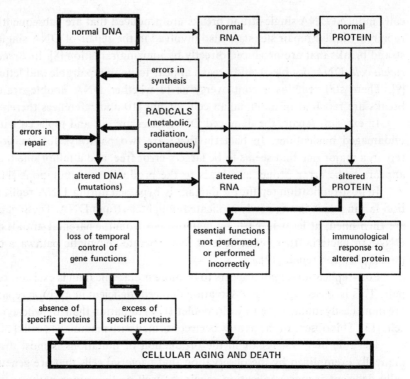

Fig. 1. Schematic representation of the genetic alteration theory of aging [19]. The diagram unifies the error, the mutation, the immunologic, the radical and the cross-linking theories of aging by focusing upon the unique importance of the fidelity of DNA to the proper functioning of a cell. The major pathways leading to cellular aging and death are indicated by the bold arrows. With very little modification, this diagram could also describe the molecular basis of carcinogenesis.

are more easily mutagenized by UV radiation and chemicals than are normal cells [14].

Workers in the field of aging have been slow to focus on the unique importance of DNA in all cellular processes, but this situation is now improving. Most of the current theories of aging, e.g. the immunologic theory, the error theory, the cross-linking theory, the radical theory, the mutation theory, are not independent theories really, but are just related aspects of what has been called the genetic alteration theory of aging [19]. Thus, the radical and cross-linking theories can be considered as descriptions of how DNA can be damaged, and the error, immunologic, and mutation theories are just descriptions of the biological consequences of damaged DNA. Figure 1 describes schematically how the various theories of aging can be combined to form a unified theory based upon alterations in the structure and function of DNA.

The genetic alteration theory presumes that the aging process results from the progressive accumulation of unrepaired or unrepairable damage to DNA due either to a decrease in the ability of cells to repair their DNA and/or to an increased propensity of the organism to damage its own DNA. This latter response could be the result of the increased metabolic production of free radicals, superoxide, hydrogen peroxide, etc. and/or a decrease in the functioning of enzymes required for the neutralization of these agents that are known to have an adverse effect on DNA.

In support of certain of these concepts, a correlation has been demonstrated between the amount of unscheduled DNA synthesis (i.e. excision repair) after the UV-irradiation of cultured cells derived from several mammalian species, and the estimated life span of these several species [20]. Thus, cells from the species with the longest life span showed the greatest ability to repair UV-induced DNA damage. In addition, there is evidence for the accumulation of DNA damage as a function of age of a given species [5, 21].

Thus, both aging and carcinogenesis appear to be related at the molecular level to damaged DNA. Such damage need not necessarily result from external agents such as radiation or chemicals, but can result from the normal metabolic production of agents that can alter DNA, e.g. radicals, hydrogen peroxide, superoxide, free electrons, etc. [22].

What the cell does or does not do to this DNA damage determines its biological effect. If the damage is repaired accurately, then it is probably of no consequence. If the damage is repaired inaccurately, then it will be mutagenic. If it is not repaired and is not lethal, as perhaps in a non-dividing cell, the damage will probably alter the expression of genes, and thus unbalance the biochemistry of the cell.

Most of our knowledge concerning the chemical nature of the damage that can be produced in DNA, its biological consequences, and the mechanisms by which it may be repaired comes from the field of radiation biology. Therefore, three fields, which at first seem philosophically divergent, actually have a strong convergence at the level of molecular mechanisms. It is encouraging, therefore, that the dialogue is improving among workers in the fields of aging, carcinogenesis, and radiation biology [6].

References

1 KAPLAN, H.S.: Biochemical basis of reproductive death in irradiated cells. Am. J. Roentg. *90:* 907–916 (1963).
2 HAYNES, R.H.: Molecular localization of radiation damage relevant to bacterial inactivation; in AUGENSTEIN, MASON and ROSENBERG Physical processes in radiation biology, pp. 51–68 (Academic Press, New York 1964).
3 WARD, J.F.: Molecular mechanisms of radiation-induced damage to nucleic acids. Adv. Radiat. Biol. *5:* 181–239 (1975).
4 VARGHESE, A.J.: Photochemistry of nucleic acids and their constituents. Photophysiology *7:* 207–274 (1972).
5 SMITH, K.C.: The radiation-induced addition of proteins and other molecules to nucleic acids; in WANG Photochemistry and photobiology of nucleic acids, pp. 187–218 (Academic Press, New York 1976).
6 SMITH, K.C. (ed.): Aging, carcinogenesis and radiation biology: the role of nucleic acid addition reactions (Plenum Publishing, New York 1976).
7 HANAWALT, P.C. and SETLOW, R.B. (eds): Molecular mechanisms for repair of DNA (Plenum Publishing, New York 1975).
8 TOWN, C.D.; SMITH, K.C., and KAPLAN, H.S.: Repair of X-ray damage to bacterial DNA. Curr. Top. Radiat. Res. Q. *8:* 351–399 (1973).
9 BONURA, T.; TOWN, C.D.; SMITH, K.C., and KAPLAN, H.S.: The influence of oxygen on the yield of DNA double-strand breaks in X-irradiated *Escherichia coli* K-12. Radiat. Res. *63:* 567–577 (1975).
10 DUGLE, D.L.; GILLESPIE, C.J., and CHAPMAN, J.D.: DNA strandbreaks, repair, and survival in x-irradiated mammalian cells. Proc. natn. Acad. Sci. USA *73:* 809–812 (1976).
11 YOUNGS, D.A.; SCHUEREN, E. VAN DER, and SMITH, K.C.: Separate branches of the *uvr* gene-dependent excision repair process in ultraviolet-irradiated *Escherichia coli* K-12 cells; their dependence upon growth medium and the *polA, recA, recB,* and *exrA* genes. J. Bact. *117:* 717–725 (1974).
12 YOUNGS, D.A. and SMITH, K.C.: Genetic control of multiple pathways of post-replicational repair in *uvrB* strains of *Escherichia coli* K-12. J. Bact. *125:* 102–110 (1976).
13 WITKIN, E.M.: Ultraviolet-light-induced mutation and DNA repair. Annu. Rev. Microbiol. *23:* 487–510 (1969).
14 MAHER, V.M. and MCCORMICK, J.J.: Effect of DNA repair on the cytotoxicity and mutagenicity of UV irradiation and of chemical carcinogens in normal and xeroderma pigmentosum cells; in YUHAS, TENNANT and REGAN Biology of radiation carcinogenesis, pp. 129–145 (Raven Press, Hewlett 1976).
15 MCCANN, J.; CHOI, E.; YAMASAKI, E., and AMES, B.N.: Detection of carcinogens as mutagens in the *Salmonella*/microsome test: assay of 300 chemicals. Proc. natn. Acad. Sci. USA *72:* 5135–5139 (1975).
16 TROSKO, J.E. and CHU, E.H.Y.: The role of DNA repair and somatic mutation in carcinogenesis. Adv. Cancer Res. *21:* 391–425 (1975).
17 CLEAVER, J.E.; BOOTSMA, D., and FRIEDBERG, E.: Human diseases with genetically altered DNA repair processes. Genetics *79:* suppl., pp. 215–225 (1975).

18 PATERSON, M.C.; SMITH, B.P.; LOHMAN, P.H.M.; ANDERSON, A.K., and FISHMAN, L.: Defective excision repair of γ-ray-damaged DNA in human (ataxia telangiectasia) fibroblasts. Nature, Lond. *260:* 444–447 (1976).
19 SMITH, K.C.: Chemical adducts to deoxyribonucleic acid: their importance to the genetic alteration theory of aging. Interdiscipl. Top. Geront. *9:* 16–24 (1976).
20 HART, R.W. and SETLOW, R.B.: Correlation between deoxyribonucleic acid excision-repair and life-span in a number of mammalian species. Proc. natn. Acad. Sci. USA *71:* 2169–2173 (1974).
21 CUTLER, R.G.: Cross-linkage hypothesis of aging: DNA adducts in chromatin as a primary aging process; in SMITH Aging, carcinogenesis and radiation biology, pp. 443–492 (Plenum Publishing, New York 1976).
22 FRIDOVICH, I.: Superoxide and evolution. Horizons Biochem. Biophys. *1:* 1–37 (1974).

Dr. K.C. SMITH, Department of Radiology, Stanford University School of Medicine, *Stanford, CA 94305* (USA)

Significance of Occupational Cancer

E. BOYLAND

London School of Hygiene and Tropical Medicine, London

Cancer must have causes and these can be biological, chemical or physical, but most cancer is caused by chemical substances. A few cases of cancer in man are due to ionising radiation from X-rays and from radioactive elements. Some skin cancer is due to sunlight in parts of the world where this is more intense than in Britain. In South Africa and Australia, however, mortality from skin cancer accounts for less than 3% of all cancers. Less than 5% of all cancer in man would appear to be caused by the physical agency of radiations.

There is no conclusive evidence that any human cancer is due to viruses although there are suggestions that some leukaemia and Burkitt's tumour may be associated with viruses. Bladder cancer in Egypt is probably caused by infection with *Schistosoma haematobium*. It seems unlikely, however, that more than 5% of the total cancer in man can be due to purely biological factors.

This means that at least 90% of human cancer is due to chemical agents. From this argument it is not possible to distinguish between environmental and endogenous chemical substances. Examination of the epidemiology of cancer indicates that about 80% of all cancer is due to environmental factors.

The first occupational cancer, cancer of the breast in nuns, was described by RAMAZZINI in 1700. At that time, he said, 'Experience attests, that where there is no placenta or impregnation, the redundant humours of the womb frequently occasion cancerous tumours in the breast; and these we meet with among nuns oftener than among other women, not from a suppression of the menses, but by reason of their living single.' RAMAZZINI's observations have been widely and generally confirmed, but the actual cause is still not known. The English translation of his book on *Diseases of Tradesman* appeared in 1705.

Table I. Scrotal cancer due to oil and tar

Author	Year	Group of workers
OGSTON	1871	paraffin workers
R. VON VOLKMANN	1875	coal distillers
JOSEPH BELL	1876	shale oil workers
S.R. WILSON	1906	mule spinners
M.D. KIPLING	1963	engineers in Birmingham
C. and J. THONY	1970	toolsetters in Savoy

In 1775, PERCIVAL POTT praised RAMAZZINI's book when he attributed the cancer of the scrotum which was common in chimney sweeps in England to the lodgement of soot in the rugae of the scrotum. This was therefore the first identification of a causative agent of cancer. Chimney sweeps cancer is now rare but scrotal cancer occurred in men working with oil and coal tar, and even today it is seen in engineering works in England and France (table I). The causes are different but the disease remains. It can, however, be prevented by good hygiene and the use of non-carcinogenic lubricating oils. The active compounds are, however, also present in polluted air.

Cancer of the bladder was identified as an occupational cancer in the chemical industry by REHN over 80 years ago. Some aromatic amines including 2-naphthylamine, 4-aminobiphenyl and benzidine are recognised as carcinogens and their use is prohibited in many countries. This has undoubtedly prevented much cancer in workers. Small amounts of naphthylamines are formed on combustion of nitrogenous material so that 2-naphthylamine is present in very low concentrations in the environment. The significance of minute amounts of carcinogens as possible human hazards is difficult to evaluate, but it is knowledge of occupational cancer which has revealed the possibility.

In spite of the evidence that chemical substances are important causes of cancer in man, the active substances are not easy to identify. Many causes of cancer in man have, however, been found by the investigation of occupational disease and this approach is still the most promising. One difficulty is that the high exposure groups are often small and the effect in such groups is difficult to see when it becomes diluted in larger groups of the population.

Another drawback to the epidemiological approach is that it is more effective in finding the causes of the rare or less common forms of the disease. The mesothelioma caused by asbestos, the nasal tumours among wood

Table II. Approximate total doses of carcinogens required to produce tumours in rodents

Carcinogen	Dose
1,2-5,6-dibenzanthracene	10 µg
Aflatoxin B	1 mg
Dimethyl nitrosamine	10 mg
Acetylaminofluorene	100 mg
Dimethylaminoazo benzene	1 g
Carbon tetrachloride	10 g

workers and the angisarcoma of the liver in men exposed to vinyl chloride are examples of such unusual tumours.

Because the actual chemicals causing cancer are so difficult to identify by examination of human populations, it is essential to carry out animal tests on chemicals in the environment to obtain indications of carcinogenic activity.

There is considerable variation in activity of carcinogenic activity of chemical compounds. Thus, some polycyclic hydrocarbons are active on injection of microgram quantities, the fungal metabolite Aflatoxin is active in milligrams when administered by feeding but the effective doses of lead acetate and carbon tetrachloride are about 10 g. Thus, there is a millionfold variation the activity of carcinogenic substances when expressed on the basis of dose required (table II).

The chemical carcinogens can be divided into groups. The first group are the directly acting substances which do not need metabolic transformation because they are already chemically reactive. These include the alkylating agents, such as nitrogen mustard, myleran and related compounds, which have been used in the chemotherapy of cancer. Mustard gas, the chemical warfare agent, and *bis*-chloromethylether, an intermediate in the production of ion exchange resins, are two compounds of this type which are known to have caused cancer in man.

Many carcinogens, however, including aromatic amines, nitrosamines, polycyclic hydrocarbons, aflatoxin, safrole and carbon tetrachloride require metabolic or chemical activation (table III). The metabolic reactions involved are part of the normal detoxication processes. Whereas the greater part of the reactive intermediates formed react with water or glutathione to form non-toxic derivatives that are easily excreted, a small proportion reacts with nucleic acids and proteins causing cell damage and genetic changes that can be expressed as mutation, embryonic damage or cancer.

Table III. Carcinogens that require metabolic activation

Aromatic amines	Aflatoxin
Nitrosamines	Safrole
Polycyclic hydrocarbons	Carbon tetrachloride
Pyrrolizidine alkaloids	4-Nitroquinoline N-oxide
Cycasin	

It is possible that these lethal biosynthetic pathways have not been prevented by evolution because cancer generally develops after reproduction at least in females is over. It is also possible that cancer is not a serious disadvantage when the survival of the fittest is concerned, because it reduces the period of old age or senility and so it would increases the proportion of active members in a population.

The first pure chemicals found to cause cancer were the polycyclic aromatic hydrocarbons. These were discovered through studies of occupational cancer in workers with oil and coal tar. The first known carcinogen 1,2-5,6-dibenzanthracene still seems to be the most active when tested by subcutaneous injection. On injection micrograms will induce sarcomas. The most widely distributed carcinogenic hydrocarbon – 3,4-benzopyrene – is formed when almost any organic matter is burned. It is present in coal tar and some mineral oils, and is probably a cause of cancer of the skin and lung in man.

Although benzopyrene and other hydrocarbons cause cancer at the site of application, they are not chemically reactive and their carcinogenic action requires metabolic change. I started work on the metabolism of aromatic compounds over 40 years ago and we now know of hundreds of metabolites. The study of the metabolism of both inactive hydrocarbons such as anthracene, naphthalene and phenanthrene as well as the carcinogenic hydrocarbons showed that the first step in the process was the formation of epoxides or arene oxides.

Such oxides formed by the enzymatic addition of oxygen to a double bond or ethylene group are reactive alkylating compounds similar to some simple epoxides which are carcinogenic. Most of the active intermediary metabolites are inactivated by detoxifying reactions including: (1) a rearrangement to give phenols; (2) reaction with water by the action of an enzyme epoxide hydrase to give diols, and (3) reaction with glutathione through the action of the enzyme glutathione epoxy transferase. The arene

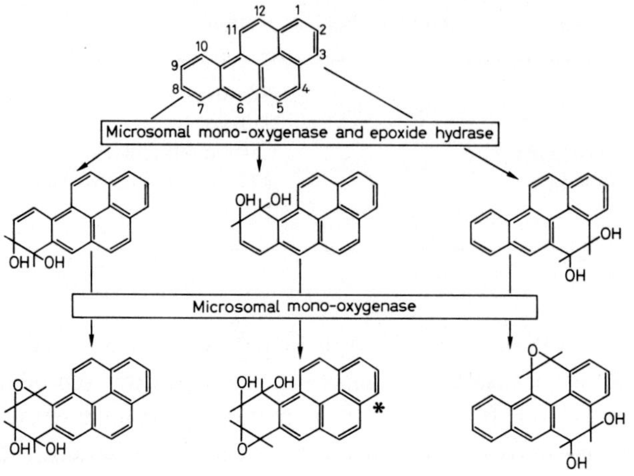

Fig. 1. Some diol-epoxides derived from benzo[a]pyrene. * = Probable proximate carcinogen (after GROVER and SIMS).

oxides can also react with macromolecules including proteins and nucleic acids. The carcinogenic hydrocarbon, 1,2,5,6-dibenzanthrene, reacts with nucleic acid more than do non-carcinogenic hydrocarbons such as phenanthrene. Carcinogenic and inactive hydrocarbons both react with protein, but the carcinogenic compounds react more rapidly with nucleic acid. This preferential reaction with nucleic acid (DNA) is probably due to the carcinogenic compounds, which are planar molecules, having more physical affinity for nucleic acid. The physical affinity could be due to the similarity in shape between the hydrocarbons and purine-pyrimidine base pairs in DNA so that they can intercalate in the helix of the DNA molecule. The reactive intermediates are probably the epoxides of dihydroxy dihydro derivatives (fig. 1).

The action of polycyclic hydrocarbons is thus due to the presence of an ethylene group or double bond which can be activated and their similarity in shape to constituents of the target molecule DNA.

Vinyl chloride which can also be called chloroethylene is a human carcinogen. This was discovered by a combination of laboratory and epidemiological studies in which the Italian workers, VIOLA and MALTONI, made the most important contributions. Research in many laboratories has shown that vinyl chloride is metabolised to an epoxide which is the reactive intermediate (table IV). This reactive molecule combines with DNA in a similar way to that of the arene oxides of carcinogenic hydrocarbons.

Table IV. Epoxides as active metabolites of aliphatic carcinogens

Vinyl chloride

$$CH_2 = CHCl \rightarrow CH_2 \overset{O}{-\!\!\!\diagup\!\!\diagdown\!\!\!-} CHCl$$

Trichloroethylene

$$CCl_2 = CHCl \rightarrow CCl_2 \overset{O}{-\!\!\!\diagup\!\!\diagdown\!\!\!-} CHCl \rightarrow CCl_3-C\!\!\diagup\!\!^{OH}_{OH}$$

A similar compound, trichloroethylene, has induced liver tumours in mice. Trichloroethylene is dichlorovinyl chloride and it was suggested some 30 years ago that trichloroethylene is metabolised through the formation of an epoxide, and this has been confirmed. One of the metabolites of trichloroethylene is chloral hydrate which can be prepared by heating trichloroethylene epoxide with alkali. This epoxide could react with DNA in the same way as the epoxides of hydrocarbons. These reactive epoxides are biologically produced alkylating agents.

These studies indicate that some ethylene derivatives or unsaturated compounds can be carcinogenic because they are metabolised to epoxides. There is little doubt that vinyl chloride, 'chloroprene' and trichloroethylene are toxic substances to which exposure should be minimal. The biochemical investigations would indicate that the saturated compounds trichloroethane and tetrachloroethane may be less hazardous than the unsaturated trichloroethylene and tetrachloroethylene as solvents. Both types of compound should be further tested for carcinogenic activity in order that carcinogenic hazards be reduced.

Occupational cancers have special significance because: (1) they can give indications of the specific causes of cancer; (2) knowledge of the specific causes can lead to their removal and so the prevention of cancer, and (3) recognition of causes of cancer in workers gives indications of causes of cancer in the general population. Thus, occupational carcinogens can be environmental carcinogens.

Prof. E. BOYLAND, London School of Hygiene and Tropical Medicine, *London WC1* (England)

Carcinogenic Bioassay of the Herbicide, 2,4,5-Trichlorophenoxyethanol (TCPE) with Different 2,3,7,8-Tetrachlorodibenzo-p-dioxin (Dioxin) Content in Swiss Mice[1]

K. Tóth, J. Sugár, Susan Somfai-Relle and Judit Bence

Research Institute of Oncopathology, Budapest

The herbicide effect of 2,4,5-trichlorophenoxyacetic acid (2,4,5-T) was recognized as early as 1944 in the USA. Since that time, 2,4,5-T and its salts and esters became widely used herbicides in other countries as well [Maier-Bode, 1972]. Due to industrial technology, the technical chlorphenols and its products like 2,4,5-T are contaminated with certain amount of chlorodibenzo-p-dioxins such as 2,3,7,8-tetrachlorodibenzo-p-dioxin (TCDD or dioxin). Dioxin has strong toxic and teratogenic properties [Gupta et al., 1973; Sparschu et al., 1971]. A high quantity of dioxin can be produced from the isomers of trichlorphenol in basic milieu at a temperature higher than 200°C [Banki, 1976].

The possible carcinogenic effect of 2,4,5-T acid and its derivatives and of dioxin is not quite clear.

In this paper, we give account of our results obtained with chronic experiments performed to study the possible carcinogenic effect of 2,4,5-trichlorophenoxyethanol (TCPE) and dioxin and to determine whether their carcinogenic effects are dose-dependent. Toxicity and carcinogenicity of dioxin administered alone were tested in a separate experimental series.

Material and Methods

In our experiments, TCPE and dioxin (Budapest Chemical Works) have been used (fig. 1). Since TCPE is almost insoluble in water, it was suspended in a salt solution containing 0.5% carboxymethyl-cellulose (L. Light & Co.

[1] The authors thank Mrs. Rose Gaal Szabo for her expert technical assistance.

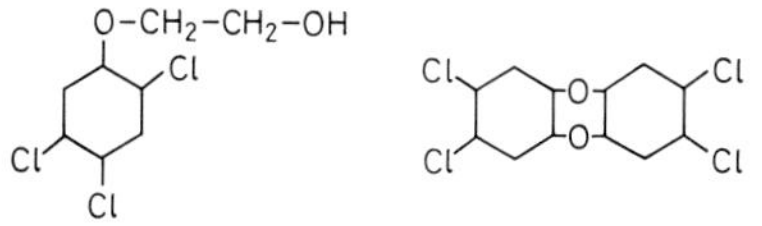

2,4,5-Trichlorophenoxy-ethanol (TCPE)   2,3,7,8-Tetrachlorodibenzo-*p*-dioxin (dioxin or TCDD)

Fig. 1. The chemical structure of TCPE and dioxin.

Table I.

Groups	TCPE mg/kg	Dioxin mg/kg	Vehicle	Present stage of experiment
1	70.0	$7 \cdot 10^{-4}$	CMC[a]	Finished after 3 years
2	70.0	$7 \cdot 10^{-6}$	CMC	
3	untreated			
4	7.0	$7 \cdot 10^{-7}$	CMC	
5	0.7	$7 \cdot 10^{-8}$	CMC	2 years (in progress)
6	7.0	$7 \cdot 10^{-5}$	CMC	
7	–	–	CMC	
8	untreated			
9	–	$7 \cdot 10^{-3}$	SO[b]	
10	–	$7 \cdot 10^{-4}$	SO	1 year (in progress)
11	–	$7 \cdot 10^{-6}$	SO	
12	–	–	SO	

[a] Carboxymethylcellulose.
[b] Sunflower oil.

Ltd). Different doses of the various admixtures of TCPE and dioxin and those of dioxin alone and vehicles were administered (table I).

Outbred Swiss-H/Riop mice were used. Each group comprised 100–100 males and females, respectively. Only the group treated with dioxin alone consisted of 50 male animals. At the onset, the mice were 10 weeks old and of 25–30 g body weight. The animals placed into plastic cages were kept at 25 ± 2°C on standard diet (Laboratory Animal Institute, Gödöllö, Hungary). Water was supplied *ad libitum*. Both compounds were administered by gastric intubation, once a week over a period of 1 year.

An acute toxicity value was determined in 6 out of 6 mice given a single intragastric dose of TCPE; death observed to the 21st day following treatment was considerable. LD_{50} value was calculated according to LICHFIELD and WILCOXON [1949].

The maximum tolerable dose in chronic treatment was determined by giving 1/10, 1/20, and 1/40 fractions of the acute LD_{50} dose to 5 males and 5 females, respectively. All the dead or moribund animals and 1 healthy mouse from each dose group were autopsied and their organs subjected to histologic examination after every 30 treatments.

During the carcinogenesis experiments, a detailed autopsy was performed mostly in dead or moribund animals and less frequently after spontaneous death; however, autopsy was always followed by histologic examination of all organs.

For the light microscope, the tissue specimens were fixed in formol, then embedded in paraffin. A routine staining procedure with haematoxylin and eosin (HE) was applied.

Amyloid deposits were identified with Congo red staining. The amyloid masses visualized in this way showed greenish birefringence under the polarizing microscope [ROMHANYI, 1943].

For the electron microscope, specimens of the liver were fixed in a 2.5% glutaraldehyde solution. Following a post-fixation in 1% osmium tetroxide, a dehydration was carried out in alcohol, then embedded in Durcupan (Fluka). The ultrathin sections were made on an LKB microtome and stained with uranyl acetate and lead citrate then studied on a JEM 6-C electron microscope.

Pathological findings of groups treated with the TCPE and/or dioxin were statistically analyzed at the end of the 2nd year of follow-up, while those of the group given dioxin alone were analyzed at the end of the 1st year. The significance was calculated by the χ_2 test.

Results

The LD_{50} value of an acute *per os* dose of TCPE contaminated with 0.1 ppm dioxin is 1,320 mg/kg. The highest dose of TCPE that did not induce any remarkable tissue lesions in the organs during 6 months of administration was 70 mg/kg; therefore, it can be regarded as the maximum tolerable dose.

The incidence of liver tumours was doubled in the males both in the first and second group treated with 70 mg/kg TCPE by the end of the 2nd year of experiments compared to that of the controls (table II). Under ex-

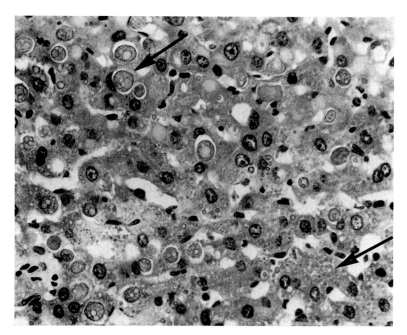

Fig. 2. Hepatoma with eosinophilic globular inclusions (arrows). HE. ×275.

Table II. Incidence of liver tumours in male mice at the end of the 2nd year of experiment

Groups	TCPE mg/kg	Dioxin mg/kg	Dioxin ppm	Vehicle	Liver tumours %
1	70.0	$7 \cdot 10^{-4}$	1.6	CMC	*43*
2	70.0	$7 \cdot 10^{-6}$	0.1	CMC	*54*
3	untreated				15–20
4	7.0	$7 \cdot 10^{-7}$	0.1	CMC	17
5	0.7	$7 \cdot 10^{-8}$	0.1	CMC	15
6	7.0	$7 \cdot 10^{-5}$	10.0	CMC	21
7				CMC	28

perimental conditions, the compound exerted a tumour-enhancing effect. However, elevated levels of dioxin concentration did not affect the tumour frequency in any of the groups during the entire period of the 3-year follow-up.

A proportionate and significant decrease of the TCPE doses (reduction to 1/10 and 1/100 of the original dose) resulted in a less frequent occurrence

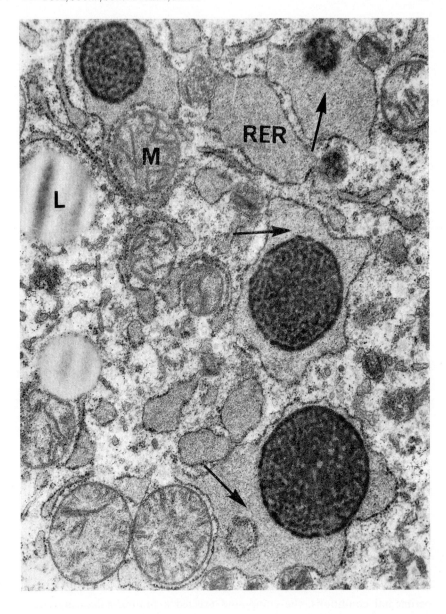

Fig. 3. Electron micrograph of a hepatoma cell. Intracisternal granules (arrows) in the dilated rough endoplasmic reticulum (RER). L = Lipid; M = mitochondrium. ×26,500.

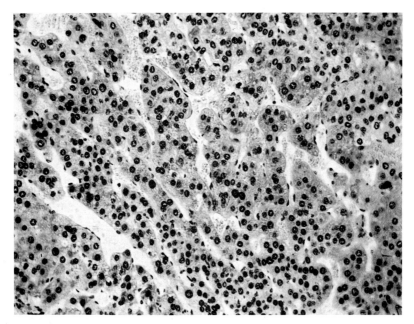

Fig. 4. Hepatocellular carcinoma of trabecular type. HE. ×170.

of the liver tumours even when a relatively higher concentration of dioxin was applied.

Tumours of the control and treated animals showed conspicuous macroscopic and microscopic similarities. Cirrhosis never preceded the development of tumours.

Histologic classification of liver tumours are as follows: (a) Hepatomas with eosinophilic globular bodies; there are no cellular atypia and metastasis formation. The cytoplasm contains several bright eosinophilic globular inclusions (fig. 2) that under the electron microscope correspond to the intracisternal granules of rough endoplasmic reticulum (fig. 3) and behave like proteins [TÓTH et al., 1975]. The fact that these intracisternal granules occurred in the tumours of both treated and control animals suggests that they are not drug-induced structures. (b) Hepatomas without inclusions; they differ from the former group so far as the globular bodies are absent. (c) Hepatocellular carcinoma of trabecular type (fig. 4). No adenomatous or anaplastic form was observed among them. These tumours are capable of forming metastases. No predominance of either histologic pattern due to the drug administration could be proved.

Table III. Pathologic effects of 1 year intragastric dioxin treatment on male Swiss mice (2 months later)

Treatment	Number of animals autopsied	Dermatitis + amyloidosis	Dermatitis only	Liver cirrhosis
7.10^{-3} mg/kg dioxin	*19*	*8*	1	3
7.10^{-4} mg/kg dioxin	4	–	–	–
7.10^{-6} mg/kg dioxin	1	–	–	–
Sunflower oil (vehicle)	1	–	–	–
Untreated (control)	50	–	4	–

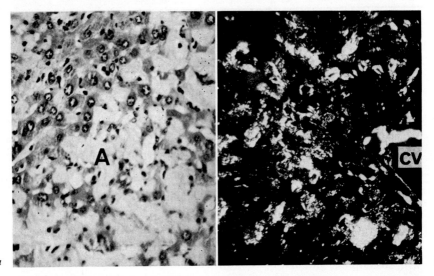

Fig. 5. (a) Amyloid deposits (A) in the liver. HE. ×250. (b) The strongly birefringent amyloid masses seen in the liver. CV = Central vein. Congo red, mounted in gum arabic, X polars, ×230.

The pathological evaluation of the groups treated with dioxin alone is considered to be preliminary as the experiments are still in progress (table III). In the group administered with 10^{-3} mg/kg dioxin, the autopsy of 19 mice revealed, in 8 cases, dermatitis associated with amyloidosis involving one or more organs (kidneys, liver, spleen). The homogeneous areas of amyloid masses stained with Congo red showed intensive birefringence (fig. 5a, b). Electron microscopy showed a fibrillar structure characteristic

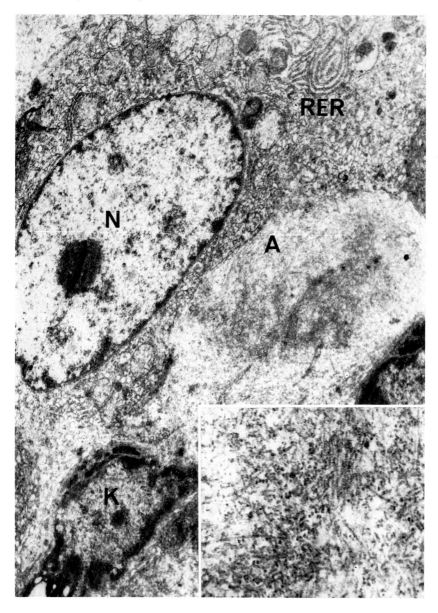

Fig. 6. Electron micrograph of amyloid deposits (A) in the liver with characteristic fibrillar structure. RER = Rough endoplasmic reticulum; N = nucleus of a hepatic cell; K = Kupffer-cell. × 3,700. Inset: The fibrils have no periodicity and branching. × 19,000.

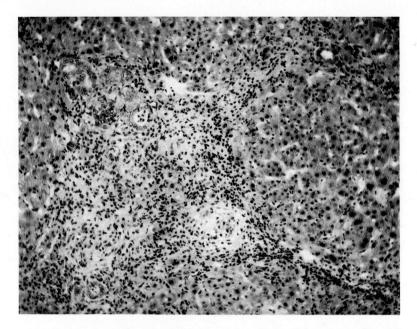

Fig. 7. Liver cirrhosis. Pseudolobuli and fibrosis with chronic inflammation. HE. ×250.

of amyloid (fig. 6). Three animals exhibited post-necrotic liver cirrhosis (fig. 7). Focal necrosis was also observed in some instances. These lesions did not develop after lower doses of the compound. No rise in the tumour incidence was experienced 2 months after the termination of the treatment.

Discussion

130 pesticides and industrial chemicals were tested for carcinogenicity in Bethesda, Md. [INNES et al., 1969]. This study included some trichlorophenol derivatives, as well, which were found to be harmless in two mouse strains. These experiments lasted for 18 months. Due to the limited period of follow-up, the results are not too convincing despite the fact that the reference carcinogens were able to induce tumours in the same strains and time allotted.

TCDD is known as one of the most toxic substances [GUPTA et al.,

1973]. Its dose-dependent teratogenic effect has been verified [SPARSCHU et al., 1971]. In humans, TCDD was found to cause chloracne, an occupational disease [BAUER et al., 1961].

The carcinogenic effect of the so-called 'less toxic' dioxin compounds have been also studied [KING et al., 1973]. In chronic experiments, high doses of these compounds resulted in severe liver lesions. The carcinogenicity cannot be evaluated for the early death of animals due to the marked toxicity of the compounds.

The authors did not examine the effects of TCDD. No data are available in the literature in regard to the possible carcinogenic effect of this molecule.

The maximum tolerable dose of TCPE (70 mg/kg) with a dioxin content of $7 \cdot 10^{-6}$ and $7 \cdot 10^{-4}$ mg/kg doubled the number of liver tumours as compared to the controls. The occurrence of liver tumours decreased when the two compounds, TCPE and dioxin, were administered in lower doses (1/10 and 1/100). The frequency of hepatomas did not change either in the first and second group (given 1.6 and 0.1 ppm dioxin) or in the sixth group that was given a relatively higher amount of dioxin (10 ppm).

A high dose of dioxin (10^{-3} mg/kg) alone caused in 3 cases post-necrotic liver cirrhosis and in 8 cases dermatitis associated with secondary amyloidosis. With lower doses, evaluable pathological lesions and a rise in tumour incidence were not observed 2 months after the termination of 1 year of treatment.

Up to now, only the hepatotoxic effect of the obviously teratogenic dioxin could be proved. The skin lesion is presumably a process analogous to the human chloracne. Amyloidosis can be regarded as a secondary process following chronic dermatitis [GLENNER and PAGE, 1976].

In the Budapest Chemical Works, TCPE is produced from imported trichlorophenol with a dioxin content of 0.1 ppm, controlled also by the factory itself. The substance is produced at low temperature and not in a basic milieu. The net dioxin content of TCPE further decreases in the course of manufacturing. The commercial product 'Buvinol' is a mixture of herbicides in which the TCPE contains dioxin only in 0.025 ppm in contrast to the maximal tolerable dose of 0.1 ppm of the WHO determination.

TCPE in the concentration of herbicides used in agriculture has no tumour-enhancing effect. The substance does not accumulate in the body but, on the contrary, becomes readily decomposed [BANKI, 1976].

The incidence of the spontaneous liver tumours is relatively high in the Swiss mouse strain used in our studies (15–20% in 3 years). However, these strains are still suitable for hepatocarcinogenicity testing because of their striking sensitivity to compounds with a tumour-enhancing effect. In our material, the increased hepatocarcinogenic effect of TCPE was noted only at the effective dose of TCPE (70 mg/kg) and, even then, only in males. From these results, one might conclude that the experimental substance (TCPE) is not a carcinogenic agent in other species since compounds inducing tumours in more species evoked tumour in either sex of Swiss mice [Tomatis et al., 1973].

We suppose that TCPE used in accordance with proper factory regulations and the plants treated with it do not mean a carcinogenic risk for the consumer.

Summary

The carcinogenic effect of a herbicide trichlorophenoxyethanol (TCPE) was investigated. The substance always contains 2,3,7,8-tetrachlorodibenzo-*p*-dioxin (dioxin) as a trace contaminant. Outbred Swiss-H/Riop mice were treated for 1 year with different doses of the various admixtures of the two compounds and those of dioxin alone to determine whether their carcinogenic effects are dose-dependent. Observation time after TCPE treatment was 1 and 2 years. The maximum tolerable dose of TCPE (70 mg/kg) with a dioxin concentration of $7 \cdot 10^{-6}$ mg/kg doubled the number of liver tumours as compared to the controls. The occurrence of liver tumours decreased at the administration of lower doses of the two compounds. No change in tumour frequency could be detected when the level of dioxin as a trace contaminant was raised, either (10 ppm). According to our preliminary results, dioxin has a more toxic than carcinogenic effect.

The use of TCPE as a herbicide in accordance with the proper factory regulations and the consumption of products treated with it does not seem to bear a carcinogenic risk for the consumer.

References

Banki, L.: The development of a herbicide, as a complex scientific task. (In Hungarian with detailed English and Russian summary.) (Medicina, Budapest 1976).
Bauer, H.; Schulz, K. und Spiegelberg, V.: Berufliche Vergiftungen bei der Herstellung von Chlorphenolverbindungen. Arch. Gewerbepath. Gewerbehyg. *18:* 538 (1961).
Glenner, G.G. and Page, D.L.: Amyloid, amyloidosis and amyloidogenesis; in Epstein and Richter International review of experimental pathology, vol. 15, pp. 2–80 (Academic Press, New York 1976).

Gupta, B.N.; Vos, J.G.; Moore, J.A.; Zinkl, J.G., and Bullock, B.C.: Pathologic effects of 2,3,7,8-tetrachlorodibenzo-*p*-dioxin in laboratory animals. Envir. Hlth Perspect. *5:* 125–140 (1973).
Innes, J.R.; Ulland, B.M.; Valerio, M.G.; Petrucelli, L.; Fischbein, L.; Hart, E.R., and Palotta, A.J.: Bioassay of pesticides and industrial chemicals for tumorigenicity in mice: a preliminary note. J. natn. Cancer Inst. *42:* 1101–1114 (1969).
King, M.E.; Shefner, A.M., and Bates, R.R.: Carcinogenesis bioassay of chlorinated dibenzodioxins and related chemicals. Envir. Hlth Perspect. *5:* 163–170 (1973).
Lichfield, J.T., jr. and Wilcoxon, F.: A simplified method of evaluating dose-effect experiments. J. Pharmac. exp. Ther. *96:* 99–113 (1949).
Maier-Bode, H.: Zur 2,4,5-T-Frage. Anz. Schädlingsk. *45:* 2–6 (1972).
Romhanyi, G.: Submikroskopische Struktur des Amyloids. Zentbl. allg. Path. path. Anat. *80:* 411 (1943).
Sparschu, G.L.; Dunn, F.L., and Rowe, V.K.: Study of the teratogenicity of 2,3,7,8-tetrachlorodibenzo-*p*-dioxin in the rat. Food Cosmet. Toxicol. *9:* 405–412 (1971).
Tomatis, L.; Partensky, G., and Montesano, R.: The predictive value of mouse liver tumour induction in carcinogenicity testing – a literature survey. Int. J. Cancer *12:* 1–20 (1973).
Tóth, K.; Somosy, Z.; Bence, J., and Sugár, J.: Study of globular bodies found in hepatomas of Swiss mice. Z. Krebsforsch. *84:* 67–73 (1975).

Dr. K. Tóth, Research Institute of Oncopathology, *Budapest* (Hungary)

Prog. biochem. Pharmacol., vol. 14, pp. 94–103 (Karger, Basel 1978)

The Problem of the Carcinogenic Risk by Furocoumarins

GIOVANNI RODIGHIERO
Istituto di Chimica farmaceutica dell'Università, Padova

Furocoumarins are a group of organic compounds in part occurring in nature, in part prepared by synthesis (fig. 1), which under irradiation with long wavelength ultraviolet light (365 nm) can exert various biological effects (table I), even at very low concentrations (μg/ml) [1,2]. Many of them are widely diffused in nature (table II); by contact of furocoumarin-containing plants or vegetable products with human skin and subsequent exposition to sunlight, various kinds of phytophotodermatitis can occur [13]. Some furocoumarins are contained in small amount in vegetables used in human nutrition, such as celery and parsley [14, 15]. Moreover, since many years they have been used for the therapeutic treatment of vitiligo [16] and recently of psoriasis [17, 18] and other cutaneous disorders of human skin [19, 20]. In

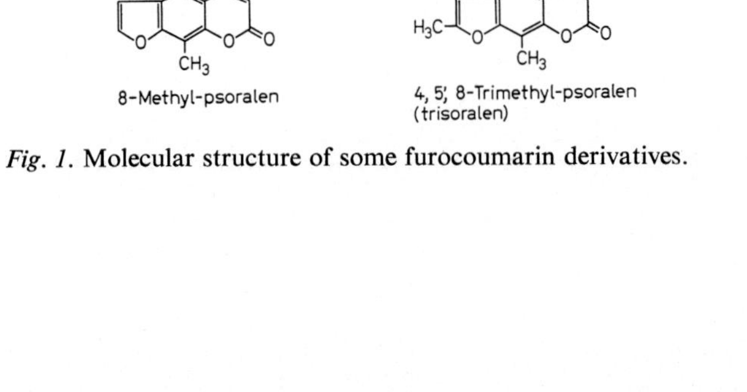

Fig. 1. Molecular structure of some furocoumarin derivatives.

Table I. Photobiological effects obtained with furocoumarins; irradiation at 365 nm

	References
Skin erythema, followed by dark pigmentation	[3]
Death of bacteria	[4, 5]
Formation of mutants in *Sarcina lutea* cultures	[6]
Mutagenic action on *Drosophyla melanogaster*	[7]
Formation of giant cells by mammalian cells adapted to *in vitro* growth	[8]
Inactivation of DNA viruses	[9]
Decrease of the template efficiency of DNA in RNA-polymerase reaction	[10]
Loss of tumor transmitting capacity of Ehrlich ascites tumor cells	[11]
Formation of polynuclear cells in sea-urchin eggs fertilized with sperm irradiated in the presence of psoralen	[12]

Table II. Occurrence of some furocoumarins in nature

Psoralen. Angelica keiskei, Coronilla glauca, C. scorpioides, Dictamus albus, Ficus carica, Heracleum antasiaticum, Phebalium argenteum, Psoralea corilifolia, P. drupacea, P. macrostachia, P. psoralioides, P. subacaulis, Ruta graveolens, R. oreojasme, R. pinnata, Zanthoxylum flavum

8-Methoxypsoralen (Xanthotoxin). Angelica archangelica, A. keiskei, Ammi majus, Dictamus albus, Fagara xanthoxyloides, Heracleum cyclocarpum, H. dissectum, H. mantegazzianum, H. sosnowskyi, Hyppomarathrum caspium, H. microcarpum, Luvunga scandens, Pastinaca opaca, P. sativa, Phebalium argenteum, Prangos fedtschenkoi, Ruta chalepensis, R. graveolens, R. montana, R. oreojasme, R. pinnata

5-Methoxypsoralen (Bergapten). Aeglopsis chevalieri, Angelica anomala, A. archangelica, A. brevicaulis, A. formosana, A. keiskei, A. pubescens, Ammi majus, Apium graveolens, Brosimum gandichandii, Citrus acida, C. bergamia, Dictamus albus, Fagara schinifolia, F. xanthoxyloides, Ficus carica, Heracleum candicans, H. cyclocarpum, H. dissectum, H. giganteum. H. lanatum, H. mantegazzianum, H. nepalense, H. panaces, H. sibiricum, H. sosnawskyi, H. sphondilium, H. verticillatum, Hippomaratrum caspium, H. microcarpum, Ligusticum acutilobum, Pastinaca opaca, P. sativa, Petroselinum sativum, Phebalium argenteum, Pheopterus littoralis, Pimpinella magna, P. saxifraga, Prangos pabularia, Ruta graveolens, R. oreojasme, R. pinnata, Seseli indicum, Severinia buxifolia, Skimmia laureola

Angelicin. Angelica archangelica, A. keiskei, A. saxicola, A. ubatakensis, Bulpleurum falcatum, Heracleum antasiaticum, H. cyclocarpum, H. dissectum, H. sosnowskyi, H. verticillatum, Psoralea coryfolia, P. drupacea, P. macrostachya

photochemotherapy, 8-methoxypsoralen (8-MOP) is orally administered and 2 h later the whole body is irradiated with long wavelength ultraviolet light (UV-A).

For these reasons, furocoumarins may incidentally come into contact with the human skin and may be introduced into the human body. Therefore, the problem of a their possible carcinogenic effect must be considered. Unfortunately, this problem at the present time has been not yet completely clarified; in fact, on one side the interactions between furocoumarins and DNA and the mutagenic activity lead one to consider them as possibly carcinogenic agents; on the other side inconclusive results have been obtained by various authors experimenting on laboratory animals and no evidence of tumor formation in man following a therapeutic treatment has been described until now.

Interactions between Furocoumarins and DNA

Two types of interaction can occur between furocoumarins and DNA.

(1) In the absence of light. Studies *in vitro* have demonstrated that a molecular complex is formed involving very weak bonds (mainly hydrophobic and Van der Waals forces). The complex exists only in aqueous solution and is not isolable; its formation has been evidenced by various ways: furocoumarins show a decrease of their UV-absorbing properties and a red shift of their absorbing maxima, while DNA shows an increase of the viscosity and a small, but well detectable, increase of its melting temperature [21, 22].

Flow-dichroism measurements demonstrated that this complex takes place by intercalation of the planar furocoumarin molecule between two base pairs in duplex DNA [23].

However, the association constant of the complex is relatively small ($1 \cdot 10^5$ for bergapten) and the highest number of furocoumarin molecules which can bind to DNA is only 2.5 every 100 nucleotides of DNA [23]. The complex, in fact, appears to be easily dissociable, for instance even by precipitation of DNA from the solution with ethyl alcohol or by gel filtration.

This type of interaction, although occurring also in living cells, does not produce any biological effect; only a very small quenching of the priming ability of DNA in DNA-dependent RNA polymerase reaction has been detected *in vitro* by the addition of furocoumarins but without irradiation

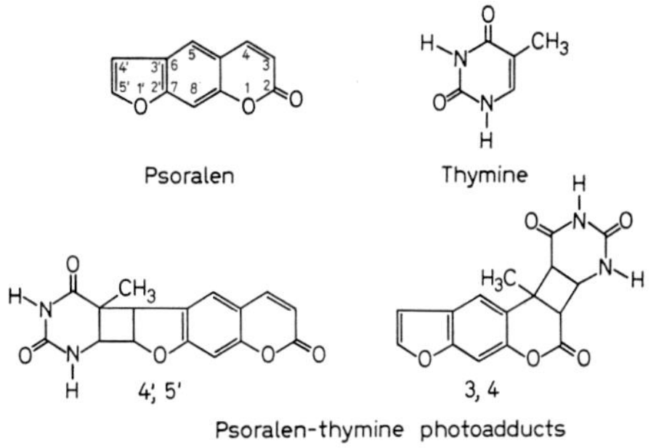

Fig. 2. Molecular structure of psoralen-thymine photoadducts.

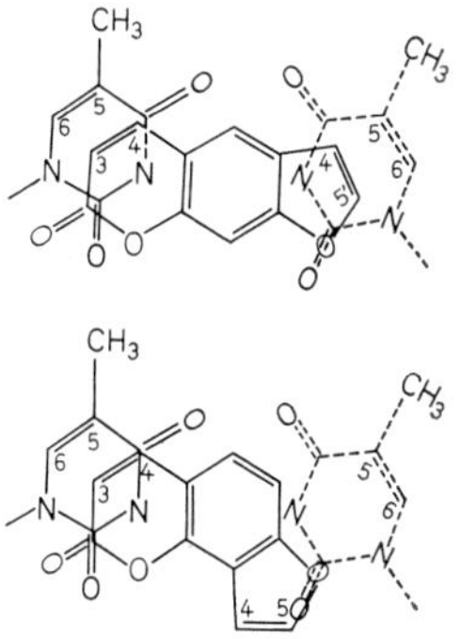

Fig. 3. Projection of the psoralen (above) and angelicin (below) molecule intercalated between two base pairs in DNA. Only two thymines are shown appertaining to the opposite strands.

[24]. No biological effects have been observed in various substrates by the simple addition of furocoumarins without irradiation.

(2) Under irradiation with long wavelength ultraviolet light (365 nm). A C_4-cyclo addition takes place between furocoumarins and pyrimidine bases, especially thymine [25], of DNA. The reactive site of pyrimidine bases is the 5,6-double bond, while furocoumarins can react either with their 3,4- or with their 4′,5′-double bond [26, 27]; therefore, two types of monoadducts can be originated (fig. 2). Furocoumarins having a linear structure (e.g. psoralen) can also photoreact with both their double bonds, forming a di-adduct with two pyrimidine bases; in native DNA this fact leads to the formation of

inter-strand cross-linkings [28, 29]. By contrast, furocoumarins having an angular structure (e.g. angelicin), for a geometric reason, cannot form cross-linkings in DNA, but only monofunctional adducts (fig. 3) [29].

Both mono- and di-functional adducts are very stable combinations in which C-C covalent linkages are involved. It was possible, in fact, to isolate the psoralen thymine mono-adducts after a complete acid hydrolysis of DNA irradiated in the presence of psoralen [30].

To this type of interaction are due all photobiological effects of furo-coumarins. As it appears from table I, besides some lethal and inhibitory effects, resulting from the destruction of the biological functions of DNA, a mutagenic action has been evidenced by various authors on different substrates. AVERBECK et al. [31], working with various furocoumarin derivatives, were able to observe that in yeast cells the damage to DNA produced by furocoumarins forming difunctional adducts leads mainly to nuclear mutations, while furocoumarins forming only mono-functional adducts lead mainly to cytoplasmic 'petite' mutations.

This mutagenic activity is one of the main indications for a possible carcinogenic action of furocoumarins.

Repair of DNA Photodamaged by Furocoumarins

A carcinogenic effect is correlated not only with the damage produced to DNA, but also with the possibility that this damage is repaired by the enzymatic systems operating in living cells. Various experiments performed both on bacterial [31–33] and cutaneous cells [34] demonstrated that the damage produced by the formation of mono-adducts is efficiently repaired

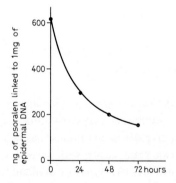

Fig. 4. Amounts of psoralen linked to the samples of DNA extracted from guinea pig skin after different periods of time following application of psoralen and irradiation at 365 nm.

(fig. 4) and DNA can recover its original biological functions. Repair of di-adducts (cross-linkages) has been extensively studied by COLE and SINDEN [35]; although possible, it appears to be less efficient in restoring the initial properties of DNA [36].

Experiments on the Carcinogenic Activity of Furocoumarins

After the application of 8-methoxy-psoralen to the therapeutic treatment of vitiligo, various authors have studied the effect of this drug on the well-known production of cutaneous tumors by UV irradiation in mice. Generally, the drug has been administered orally or by intraperitoneal injection or has been applied on the skin; mice have been then irradiated for a few minutes a day, continuing the treatment for several weeks. The results obtained by various authors are summarized in table III. As it appears, the results obtained are strongly dependent both from the way of introduction of the drug and from the wavelength of the radiation used. Although some conflicting results are present, it seems possible to conclude that epicutaneous application of furocoumarins and irradiation with long-wavelength UV light has the consequence to produce cutaneous tumors. By contrast, oral administration of the drug seems to protect from the carcinogenesis due to irradiation with short wavelength UV radiation and to have no carcinogenic effect by irradiation with long wavelength UV radiation. The reason for this different behaviour is not clear at the present time; this is an important problem and therefore further research must be done for its clarification.

Concerning the possible carcinogenesis in man, it must be emphasized that no evidence of cancer formation has been till now described after a

Table III. Effect of 8-MOP on the ultraviolet carcinogenesis in mice

O'NEAL and GRIFFIN [37] – Mercury arc lamp; orally; protection; i.p.: increase
GRIFFIN [38] – Lamp emitting mainly at 2,537 Å; orally; total protection; i.p.: no effect. Lamp emitting mainly at 3,655 Å; orally; small increase; i.p.: increase
URBACH [39] – Lamp emitting mainly at 3,100 Å; orally; no effect; i.p.: increase; topically: increase
PATHAK *et al.* [40] – Mercury arc lamp; orally; no effect; i.p.: no effect
URBACH *et al.* [41] – Xenon arc solar simulator (spectrum very close to that of sunlight); topically: increase
PATHAK [42] – 365 nm radiation; orally; no effect; topically: tumor formation
FRY *et al.* [43] – 365 nm radiation; topically: tumor formation

Table IV. Elements for a possible photocarcinogenic activity of furocoumarins

Interaction with DNA: positive
Mutagenic activity: positive
Experiments on laboratory animals:
   (a) By epicutaneous application: positive
   (b) By oral administration: negative
Evidence for tumor formation in man: negative

therapeutic treatment of vitiligo and psoriasis. This fact may signify that furocoumarins have not a carcinogenic activity in man. However, since this conclusion appears to be in contrast with the theoretical prevision due to the biological properties of furocoumarins and a tumor may appear even many years after the treatment, very careful, wide and prolonged clinical observations are still necessary prior to drawing a sure conclusion.

Final Remarks

Although the problem of the carcinogenic activity of furocoumarins is at the present still open (table IV) and further research is necessary for clarifying some conflicting aspects, some remarks and partial conclusions can be drawn from what we have till now exposed about the properties of furocoumarins.

We must consider that a carcinogenic effect may eventually occur only after irradiation; in fact, as we have seen, the dark interaction (I type) of furocoumarins with DNA has no biological consequences. Therefore, although furocoumarins are administered orally in the photochemotherapy of vitiligo and psoriasis and therefore are distributed in all parts of the body, we must expect that a carcinogenic effect may occur eventually only in the skin, because only the skin can be exposed to UV radiation and the photoreaction with the pyrimidine bases of DNA can only take place on the skin. This is a very important conclusion which is surely evident from the preceding considerations.

A second remark derives from considerations about the repair mechanisms; although the repair of damage to DNA produced by furocoumarins has not yet been studied in human skin (it has been studied however in guinea pig skin [34], we can assume that it must be very efficient. However, we know that the effectiveness of the repair is high when the damage is small. Therefore, in the therapeutic treatments, great care must be taken that the damage to DNA is as small as possible, avoiding overdosages both of the drug and of the radiation.

References

1. MUSAJO, L. and RODIGHIERO, G.: The skin-photosensitizing furocoumarins. Experientia *18:* 153–161 (1962).
2. MUSAJO, L. and RODIGHIERO, G.: Mode of photosensitizing action of furocoumarins; in GIESE Photophysiology, vol. 7, p. 115 (Academic Press, New York 1972).
3. KUSKE, H.: Experimentelle Untersuchungen zur Photosensibilisierung der Haut durch pflanzliche Wirkstoffe. Arch. Derm. Syph. *178:* 112–123 (1938).
4. FOWLKS, W.L.; GRIFFITH, D.G., and OGINSKY, E.L.: Photosensitization of bacteria by furocoumarins and related compounds. Nature, Lond. *181:* 571–572 (1958).
5. OGINSKY, E.L.; GREEN, G.S.; GRIFFITH, D.G., and FOWLKS, W.L.: Lethal photosensitization of bacteria with 8-methoxy-psoralen to long wavelength ultraviolet radiation. J. Bact. *78:* 821–833 (1959).
6. MATHEWS, M.M.: Comparative studies of lethal photosensitization of *Sarcina lutea* by 8-methoxy-psoralen and by toluidine blue. J. Bact. *85:* 322–331 (1963).
7. NICOLETTI, B. e TRIPPA, G.: Sull'azione mutagena del psoralene irradiato con luce ultravioletta in *Drosophyla melanogaster*. Rend. Accad. Naz. Lincei, Roma *43:* 259–263 (1967).
8. COLOMBO, G.; LEVIS, A.G., and TORLONE, V.: Photosensitization of mammalian cells and of animal viruses by furocoumarins. Prog. biochem. Pharmacol., vol. 1, pp. 392–399 (Karger, Basel 1965).
9. MUSAJO, L.; RODIGHIERO, G.; COLOMBO, G.; TORLONE, V., and DALL'ACQUA, F.: Photosensitizing furocoumarins: interactions with DNA and photoinactivation of DNA containing viruses. Experientia *21:* 24–25 (1965).
10. CHANDRA, P. and WACKER, A.: Photodynamic effects on the template activity of nucleic acids. Z. Naturforsch. *21b:* 663–666 (1968).
11. MUSAJO, L.; VISENTINI, P.; BACCICHETTI, F., and RAZZI, M.A.: Photoinactivation of Ehrlich ascites tumor cells *in vitro* obtained with skin-photosensitizing furocoumarins. Experientia *23:* 335–336 (1967).
12. COLOMBO, G.: Photosensitization of sea-urchin sperm to long wavelength ultraviolet light by psoralen. Exp. Cell Res. *48:* 167–169 (1967).
13. PATHAK, M.A.: Phytophotodermatitis; in PATHAK, HARBER, SEIJI and KUKITA Sunlight and man, p. 495 (University of Tokyo Press, Tokyo 1974).
14. MUSAJO, L.; CAPORALE, G. e RODIGHIERO, G.: Isolamento del bergaptene dal sedano e dal prezzemolo. Gazz. Chim. Ital. *84:* 870–873 (1954).
15. RODIGHIERO, G. e ALLEGRI, G.: Ricerche sul contenuto in bergaptene del sedano e del prezzemolo. Farmaco *14:* 727–733 (1959).
16. FITZPATRICK, T.B.; PARRISH, J.A., and PATHAK, M.A.: Phototherapy of vitiligo

(idiopathic leukoderma); in Pathak, Harber, Seiji and Kukita Sunlight and man, pp. 783–791 (University of Tokyo Press, Tokyo 1974).

17   Parrish, J.A.; Fitzpatrick, T.B.; Tanenbaum, L., and Pathak, M.A.: Photochemotherapy of psoriasis with oral methoxsalen and longwave ultraviolet light. New Engl. J. Med. *291:* 1207–1222 (1974).

18   Wolff, K.; Hönigsmann, H.; Gschnait, F. und Konrad, K.: Photochemotherapie der Psoriasis; in Jung Photochemotherapie, pp. 81–89 (Schattauer, Stuttgart 1975).

19   Hönigsmann, H.; Konrad, K.; Gschnait, F., and Wolff, K.: Photochemotherapy of mycosis fungoides. Comm. 7th Int. Congr. Photobiology, Rome 1976.

20   Uematzu, S. and Mizuno, N.: Methoxalen photochemotherapy of pustulosis palmaris and plantaris. Comm. 7th Int. Congr. Photobiology, Rome 1976.

21   Dall'Acqua, F. and Rodighiero, G.: The dark-interaction between furocoumarins and nucleic acids. Rend. Accad. Naz. Lincei, Rome *40:* 411–422 (1966).

22   Dall'Acqua, F. and Rodighiero, G.: Changes in the melting curve of DNA after the photoreaction with skin-photosensitizing furocoumarins. Rend. Accad. Naz. Lincei, Rome *40:* 595–600 (1966).

23   Dall'Acqua, F.: New chemical aspects of the photoreaction between furocoumarins and DNA. Lecture 7th Int. Congr. Photobiology, Rome 1976.

24   Chandra, P.; Marciani, S.; Dall'Acqua, F.; Vedaldi, D.; Rodighiero, G., and Biswas, R.K.: Structure specificity of polydeoxyribonucleotides for the photoreaction with psoralen. FEBS Lett. *35:* 243–246 (1973).

25   Marciani, S.; Dall'Acqua, F.; Vedaldi, D., and Rodighiero, G.: Receptor sites of DNA for the photoreaction with psoralen. Farmaco *31:* 140–151 (1976).

26   Musajo, L.; Bordin, F.; Caporale, G.; Marciani, S., and Rigatti, G.: Photoreactions at 3655 Å between pyrimidine bases and skinphotosensitizing furocoumarins. Photochem. Photobiol. *6:* 711–719 (1967).

27   Musajo, L.; Bordin, F., and Bevilacqua, R.: Photoreactions at 3,655 Å linking the 3,4-double bond of furocoumarins with pyrimidine bases. Photochem. Photobiol. *6:* 927–931 (1967).

28   Dall'Acqua, F.; Marciani, S., and Rodighiero, G.: Inter-strand cross-linkages occurring in the photoreactions between psoralen and DNA. FEBS Lett. *9:* 121–123 (1970).

29   Dall'Acqua, F.; Marciani, S.; Ciavatta, L., and Rodighiero. G.: Formation of inter-strand cross-linkings in the photoreactions between furocoumarins and DNA. Z. Naturforsch. *26b:* 561–569 (1971).

30   Musajo, L.; Rodighiero, G.; Dall'Acqua, F.; Bordin, F.; Marciani, S. e Bevilacqua, R.: Prodotti di fotocicloaddizione a basi pirimidiniche isolati da DNA idrolizzato dopo irradiazione a 3,655 Å in presenza di psoralene. Rend. Accad. Naz. Lincei, Roma *42:* 457–468 (1967).

31   Averbeck, D.; Biswas, R.K., and Chandra, P.: Photoinduced mutations by psoralens in yeast cells; in Jung Photochemotherapie, pp. 97–104 (Schattauer, Stuttgart 1975).

32   Chandra, P.; Rodighiero, G.; Dall'Acqua, F.; Marciani, S.; Kraft, S., and Wacker, A.: Studies on the reactivation of bacteria photodamaged by furocoumarins. Studia biophysica *29:* 53–61 (1971).

33   Chandra, P.; Dall'Acqua, F.; Marciani, S., and Rodighiero, G.: Studies on the repair of DNA photodamaged by furocoumarins; in Pathak, Harber, Seiji and

KUKITA Sunlight and man, pp. 411–417 (University of Tokyo Press, Tokyo 1974).
34 DALL'ACQUA, F.; MARCIANI, S.; VEDALDI, D., and RODIGHIERO, G.: Formation of inter-strand cross-linkings in DNA of guinea pig skin after application of psoralen and irradiation at 365 nm. FEBS Lett. *27:* 192–194 (1972).
35 COLE, R.S. and SINDEN, R.R.: Psoralen cross-links in DNA: biological consequences and cellular repair; in NYGAARD, ADLER and SINCLAIR Radiation research, pp. 582–589 (Academic Press, New York 1974).
36 BORDIN, F.; CARLASSARE, F.; BACCICHETTI, F., and ANSELMO, L.: DNA repair and recovery in *Escherichia coli* after psoralen and angelicin photosensitization. Biochim. biophys. Acta *447:* 249–259 (1976).
37 O'NEAL, M.A. and GRIFFIN, A.C.: The effect of oxypsoralen upon ultraviolet carcinogenesis in albino mice. Cancer Res. *17:* 911–916 (1957).
38 GRIFFIN, A.C.: Methoxsalen in ultraviolet carcinogenesis in the mouse. J. invest. Derm. *32:* 367–372 (1959).
39 URBACH, F.: Modification of ultraviolet carcinogenesis by photoactive agents. J. invest. Derm. *32:* 373–378 (1959).
40 PATHAK, M.A.; DANIELS, F.; HOPKINS, C.E., and FITZPATRICK, T.B.: Ultraviolet carcinogenesis in albino and pigmented mice receiving furocoumarins: psoralen and 8-methoxy-psoralen. Nature, Lond. *183:* 728–730 (1959).
41 URBACH, F.; FORBES, P.D., and DAVIES, R.E.: On the relationship of phototoxicity to photocarcinogenesis; in JUNG Photochemotherapie pp. 115–118 (Schattauer, Stuttgart 1975).
42 PATHAK, M.A.: Personal commun.
43 FRY, R.J.M.; GRUBE, D.D., and LEY, R.D.: The tumorigenic effects of exposure to 8-methoxy-psoralen and u.v. light. Comm. 7th Int. Congr. on Photobiology, Rome 1976.

Dr. G. RODIGHIERO, Istituto di Chimica farmaceutica dell'Università, *Padova* (Italy)

Perinatal Viral Infections and the Risk of Certain Cancers

N. Muñoz

International Agency for Research on Cancer, Lyon

A close link between teratogenesis and carcinogenesis is suggested by human and experimental data. In humans, an association between certain congenital malformations and certain cancers has been reported [1], for example, an increased risk for leukaemia has been observed in patients with Down's syndrome, Bloom's syndrome and Fanconi's anaemia. Prenatal exposure to stilboestrol has been associated with an increased risk for vaginal and cervical adenocarcinoma [2] and with an increased risk for congenital malformations of the genital tract in the female offspring of the exposed women [3]. Experimental studies have demonstrated that the critical factor which determines the ability of a chemical to produce congenital malformations or tumours following prenatal exposure is the timing of its administration during pregnancy. A chemical has an embryotoxic effect when it is administered very early during pregnancy, a teratogenic effect if it is administered later in the first half of pregnancy, and a carcinogenic effect when it is administered during the second half of pregnancy [4].

If time of exposure in pregnancy is the factor which determines whether a given chemical acts as teratogen or carcinogen, the same could be valid for viruses. This possibility becomes even more interesting if we consider that there are several viruses with recognized or suspected teratogenic properties. These viruses are: rubella and cytomegalovirus which are responsible for most of the malformations of viral origin; herpes simplex and varicella which also are teratogenic but which appear to be less important causes of malformations, and influenza, mumps and coxsackie viruses which are suspected teratogens [1–5]. It is then possible that certain viral infections during the perinatal period might play a role in the development of certain tumours. The evidence for this possibility will be reviewed.

Influenza Virus

Several epidemiological studies which suggest an association between prenatal influenza and childhood cancer, especially leukaemia, have been reported [5–9]. However, studies which do not show such an association have also been reported [9–11]. Several reasons could be advocated to explain this inconsistency: in most studies population exposure to influenza instead of individual exposure was used; in the few studies in which individual exposure to influenza was used, such exposure was ascertained by interview and not serologically; the size of the populations studied and the different approaches to the analysis of the data could also explain the controversial results [12].

There is also evidence which suggests that influenza virus may have teratogenic properties, but confirmation of this effect is lacking [13].

Varicella Virus

Chickenpox during pregnancy has also been associated with an increased cancer risk in the offspring. In a cohort study of 272 children born to mothers who had chickenpox during pregnancy, an increased risk for leukaemia was reported [14]. The same association for leukaemia and medulloblastoma was observed in a retrospective study [6]. Recently, a good correlation between the seasonal variations, the urban-rural differences and the time trends of varicella, and those of acute leukaemia, has been reported [15]. The same authors carried out a retrospective search and found 3 cases of leukaemia among the offspring of 63 women who had had varicella during pregnancy, while much less than 1 case was expected [15]. Although these results are suggestive, the numbers involved in three studies are too small to draw any firm conclusion on the role of chickenpox in these tumours.

Varicella virus appears to have teratogenic effects, but it is a relatively uncommon cause of congenital malformations [16].

SV-40

An increased risk for cancer, and especially for tumours of neural origin, has been reported in the offspring of women who were immunized with killed polio vaccine contaminated with SV-40 [17].

The data for this study were derived from the US Collaborative Perinatal Project. A total of 50,897 pregnancies were studied from 12 hospitals from 1959 to 1965. Of these, 18,342 women were immunized during pregnancy with killed polio vaccine, which later on was found to be contaminated with SV-40, and 32,555 women were not immunized. The children born to these two groups of women were similar in relation to maternal age, race, socio-economic level, maternal exposures to drugs and X-rays and in the quality of the follow-up. 14 cancers occurred among the children born to immunized mothers ($7.6 \times 10,000$) and 10 among the children of non-immunized mothers ($3.1 \times 10,000$). The cancer rate among the exposed children was even higher ($15 \times 10,000$) when the vaccine was given during the first trimester. Seven out of the 14 tumours in the exposed children were of neural origin which represents a rate ten times higher in the exposed children as compared with the non-exposed children [17].

Although these data are suggestive of an association between SV-40 and certain childhood cancers, especially if we consider that SV-40 is oncogenic for several laboratory animals and it is able to transform human cells *in vitro*, confirmation of these findings is needed before a causal association could be established.

Hepatitis B Virus

There is data which suggest that intrauterine or perinatal infection with hepatitis B virus (HBV) might be one of the risk determinants for a cancer in adults, namely liver cell cancer. Seroepidemiological studies have shown that although infection with this virus is widespread, it is more common in developing countries and especially in countries with high incidence of liver cancer, such as African and Asian countries [18]. Antigens and antibodies to the virus have been found in higher proportion among patients with liver cancer than among controls from both high and low risk areas for liver cancer, but the most striking difference between cases and controls and between healthy populations from high and low risk areas for liver cancer is in the prevalence rate of carriers of the hepatitis B surface antigen (HBsAg). The carrier rate is higher among liver cancer patients (40–80%) than among controls (0–15%) and it is also higher among the healthy population of high risk countries for liver cancer (10–15%) than among the healthy population of low risk countries (0–1%) [18–20]. In addition, it has been suggested that perinatal infection with the virus accounts for most of the carriers in some countries [21]. Recent studies have shown a very high proportion of carriers

(80–90%) among mothers of liver cancer patients from Senegal [22]. This indicates that these mothers were presumably carriers of the antigen during the perinatal period of the liver cancer patients. All these data suggest that perinatal infections with the HBV develop more readily into a chronic persistent hepatitis and that this chronic persistent infection is a high risk condition for liver cancer. A cohort study of carriers and non-carriers is needed to strengthen the association between HBV and liver cancer. If a causal relationship is established, we must keep in mind that HBV might be only one of the many factors involved in the development of liver cancer. Among the other factors or co-factors, aflatoxins appear to be of prime importance, at least in African and Asian countries. To study the combined effect of HBV and aflatoxin in the development of liver cancer, the IARC is planning to co-ordinate a series of cohort studies of carriers and non-carriers of HBV in areas with presumably varying exposure to aflatoxin.

In summary, although there is epidemiological evidence which suggests that perinatal viral infections might be one of the risk determinants of certain childhood and adult cancers, confirmation of this association will come from prospective studies in which individual exposure to the virus in question is ascertained serologically. Since cancer in children is a rare disease, collaborative studies in which data from various cohorts are pooled are necessary to study this problem. The IARC is co-ordinating a series of prospective studies which were, or are, being carried out to study the role of prenatal events on congenital malformations and to study also the role of the same intrauterine exposures in the development of cancer in children. Intrauterine exposures to radiation, acute and chronic diseases, medicaments and occupational exposures will be analyzed. Serological confirmation of viral diseases will be possible in those studies in which serum samples were, or are, being collected during pregnancy. A cohort of approximately 120,000 pregnancies have been assembled from the various European studies. A total of 64 cancer cases are expected to develop among children born to these pregnancies. Identification of these children is under way.

References

1. MILLER, R.W.: Relation between cancer and congenital defects: an epidemiological evaluation. J. natn. Cancer Inst. *40:* 1079–1085 (1968).
2. HERBST, A.L., et al.: Clear cell adenocarcinoma of the vagina and cervix in girls: analysis of 170 registry cases. Am. J. Obstet. Gynec. *119:* 713–724 (1974).
3. HERBST, A.L., et al.: Prenatal exposure to stilboestrol. A prospective comparison of ex-

posed female offspring with unexposed controls. New Engl. J. Med. *292:* 334–339 (1975).
4   TOMATIS, L.: Transplacental carcinogenesis; in RAVEN Modern trends in oncology, p. 99 (Butterworth, London 1973).
5   FEDRICK, J. and ALBERMAN, E.D.: Reported influenza in pregnancy and subsequent cancer in the child. Br. med. J. *ii:* 485–488 (1972).
6   BITHELL, J.F.; DRAPER, G.J., and GORBACH, P.D.: Association between malignant disease in children and maternal virus infections. Br. med. J. *i:* 706–708 (1973).
7   HAKULINEN, T., *et al.:* Association between influenza during pregnancy and childhood leukemia. Br. med. J. *iv:* 265–267 (1973).
8   AUSTIN, D.F., *et al.:* Excess leukemia in cohorts of children born following influenza epidemics. Am. J. Epidem. *101:* 77–83 (1975).
9   RANDOLPH, V.C. and HEATH, C.S.: Influenza during pregnancy in relation to subsequent childhood leukemia and lymphoma. Am. J. Epidem. *100:* 399–409 (1974).
10 LECK, I. and STEWARD, J.K.: Incidence of neoplasms in children born after influenza epidemics. Br. med. J. *iv:* 631–634 (1972).
11 MCCREA CURNEN, M.G. *et al.:* Childhood leukemia and maternal infectious diseases during pregnancy. J. natn. Cancer Inst. *53:* 943–947 (1974).
12 MUÑOZ, N.: Prenatal exposure and carcinogenesis; in EMANUELLI Proc. 16th Postgrad. Course on Malignant Tumours in Children, pp. 27–30 (Ambrosiana, Milan 1975).
13 MACKENZIE, J.S. and HOUGHTON, M.: Influenza infections during pregnancy: association with congenital malformations and with subsequent neoplasms in children, and potential hazards of live virus vaccines. Bact. Rev. *38:* 356–370 (1974).
14 ADELSTEIN, A.M. and DONOVAN, J.W.: Malignant disease in children whose mothers had chickenpox, mumps or rubella in pregnancy. Br. med. J. *iv:* 629–631 (1972).
15 VIANNA, N. and POLAN, A.: Childhood lymphatic leukemia: prenatal seasonality and possible association with congenital varicella. Am. J. Epidem. *103:* 321–332 (1976).
16 DUDGEON, J.A.: Infective causes of human malformations. Br. med. Bull. *32:* 77–83 (1976).
17 HEINONEN, O.P., *et al.:* Immunization during pregnancy against poliomyelitis and influenza in relation to childhood malignancy. Int. J. Epidem. *2:* 229–235 (1973).
18 NISHIOKA, K.; LEVIN, A.G., and SIMONS, M.J.: Hepatitis B antigen, antigen subtypes, and hepatitis B antibody in normal subjects and patients with liver disease. WHO Bull. *52:* 293–300 (1975).
19 PRINCE, A.M.; SZMUNESS, W.; MICHON, J.; DEMAILLE, J.; DIEBOLT, G.; LINHARD, J.; QUENUM, C., and SANKALÉ, M.: A case/control study of the association between primary liver cancer and hepatitis B infection in Senegal. Int. J. Cancer *16:* 376–383 (175).
20 MAUPAS, P., *et al.:* Antibody to hepatitis B core antigen in patients with primary hepatic carcinoma. Lancet *July:* 9–11 (1975).
21 SCHWEITZER, I.L.: Vertical transmission of the hepatitis B surface antigen; in Proc. Symp. on Viral Hepatitis, pp. 287–291 (Slack, Thorofare 1975).
22 BLUMBERG, B.S.; LAROUZÉ, B.; LONDON, W.T.; WERNER, B.; HESSER, J.E.; MILLMAN, I.; SAIMOT, G., and PAYET, M.: The relation of infection with the hepatitis B agent to primary hepatic carcinoma. Am. J. Path. *81:* 669–682 (1975).

Dr. N. MUÑOZ, International Agency for Research on Cancer, 150, Cours Albert Thomas, *F-69008 Lyon* (France)

Immunology of Rat Hepatic Neoplasia[1]

R.W. BALDWIN

Cancer Research Campaign Laboratories, University of Nottingham, Nottingham

Several distinct types of neoantigen may be expressed on carcinogen-induced tumours of the rat liver [BALDWIN, 1973; PRICE and BALDWIN, 1977]. These include tumour rejection antigens and cell surface antigens, both of which are characteristic for individual tumours, as well as cross-reacting fetal antigens. In addition a variety of 'abnormal' antigens may be demonstrated, sometimes transiently, during early stages of hepatocarcinogenesis, these being normal liver components modified by covalent interaction with carcinogen metabolites [BALDWIN, 1962; KITAGAWA et al., 1966]. Whilst it is unlikely that all of these neoantigens play a significant role in host immunosurveillance [BALDWIN, 1976], they may be viewed as specific markers for characterizing transformed cells and for studying metabolic events during carcinogenesis.

Tumour-Specific Rejection Antigens

Tumour-specific rejection antigens are generally identified by their capacity to elicit immune rejection responses against tumour cells transplanted in syngeneic hosts. This may be achieved, for example, by surgical resection of a developing tumour graft or implantation of tumour cells attenuated by X- or γ-irradiation [BALDWIN and BARKER, 1967]. Immunity may also be induced by implantation of viable tumour cells in admixture with immunological adjuvants such as bacillus Calmette Guérin (BCG) [BALDWIN and PIMM, 1976]. This often results in rejection of the transplanted tumour and produces an enhanced immunological response.

[1] Supported by a grant from the Cancer Research Campaign.

Employing these procedures, hepatic tumours induced by feeding 4-dimethylaminoazobenzene (DAB) and 3'-methyl-DAB have been shown to express tumour-associated rejection antigens [GORDON, 1965; BALDWIN and BARKER, 1967]. This is exemplified by tests with a group of DAB-induced hepatic neoplasms in Wistar rats where immunization either by tumour excision or implantation of γ-irradiated (15,000 R) tumour consistently induced tumour immunity, treated rats rejecting challenge with up to 5×10^5 viable cells of the immunizing tumour. So far, there has been no attempt to systematically characterize the immunogenicities of hepatic neoplasms induced with other aminoazo dye carcinogens although 5-(ϱ-dimethylamino-phenyl-azo) quinoline-induced tumours also proved to be active [BALDWIN et al., 1973a].

Hepatic tumours induced in rats with diethylnitrosamine (DENA) are also immunogenic as defined by the capacity of γ-irradiated tumour cells to elicit rejection of transplanted tumour in syngeneic recipients [BALDWIN and EMBLETON, 1971a]. In contrast to these findings with aminoazo dye- and DENA-induced hepatic tumours, those induced by 2-acetylaminofluorene (AAF) express little or no immunogenicity [BALDWIN and EMBLETON, 1969]. Thus, tumour rejection reactions could only be detected with 3/10 hepatic tumours and, with these, the level of immunity, reflected by the maximum tumour challenge rejected by immunized rats, was low. It is notable that other tumours arising in AAF-treated rats, especially mammary carcinomas, also lack significant immunogenicity [BALDWIN and EMBLETON, 1969, 1974]. No satisfactory explanation for these observed differences in the immunogenicities of hepatic neoplasms induced by different chemical carcinogens has yet been provided [cf. BALDWIN, 1973], but these observations indicate that the expression of tumour-associated rejection antigens is not a necessary concomitant of neoplastic transformation.

One early premise that the development of non-immunogenic tumours reflected immunoselective effects in the tumour-bearing host developed from several studies where tumour immunogenicity showed a degree of correlation with the latent induction period. It was found by BARTLETT [1972], for example, that early arising 3-methylcholanthrene (MCA)-induced murine sarcomas displayed a wide range of immunogenicities, whereas with progressively increasing latent periods of induction, the proportion of highly immunogenic tumours in the population decreased. Similarly, in the rat tumour systems already referred to, the immunogenic DAB-induced hepatomas arose much earlier than the weakly or non-immunogenic AAF-induced tumours [BALDWIN and EMBLETON, 1969]. This hypothesis has received little support, however, since there is no conclusive evidence in favour of immunosurveillance against chemically induced tumours [BALDWIN, 1973, 1976; STUTMAN, 1975]. Furthermore, it has been shown that tumours induced under conditions where immunosurveillance cannot be operative also exhibit variability in their expression of tumour-specific rejection antigens. For example, mouse prostate cells transformed *in vitro* by polycyclic hydrocarbons show variability in the expression of tumour-specific rejection antigens as defined by their capacity to elicit immunity to tumours developing *in vivo* [MONDAL et al., 1970]. Similar variability of tumour specific antigen expression could also be demonstrated by comparing *in vitro* reactions of immune sera or sensitized lymphoid cells from syngeneic mice, immunized with these transformed cells [EMBLETON and HEIDELBERGER, 1972].

Another hypothesis, not as yet adequately evaluated, is that the expression of tumour-specific antigen reflects either qualitatively or quantitatively carcinogen-induced derangement within the transformed cell. For example, PREHN [1975] has reported that the immunogenicity of MCA-induced murine sarcomas is proportional to the dose of carcinogen administered, low-dose carcinogenic stimuli producing tumours of low immunogenicity.

It is more difficult to compare responses to different carcinogens, especially when administered orally, but in the studies on rat hepatic tumours the doses of DAB administered are greater than those of AAF [BALDWIN and EMBLETON, 1969, 1971a]. Consequently, it is feasible that the differences in the immunogenicity of tumours induced by AAF compared with other carcinogens may reflect dose responses and are not due to qualitative effects induced by different classes of carcinogen.

Undoubtedly, the most significant feature of the tumour rejection antigens associated with chemically induced hepatic tumours is their great diversity and at the present time there are few exceptions to the general finding that each tumour has a characteristic antigen [BALDWIN, 1973]. This was established with DAB-induced rat hepatic tumours in tests showing that immunization only elicited resistance to challenge with cells of the immunizing tumour [BALDWIN and BARKER, 1967]. This specificity is further illustrated by experiments showing that four distinct hepatic nodules arising in a rat treated with 3'-methyl-DAB were antigenically distinct [ISHIDATE, 1970].

Tumour-Specific Cell Surface Antigens

Neoantigens expressed at the cell surface of carcinogen-induced rat hepatic neoplasms have been identified by *in vitro* analyses of cell-mediated and humoral immunity elicited against tumours transplanted into syngeneic recipients [BALDWIN, 1973]. These *in vitro* reactions have closely correlated with the expression of tumour rejection antigens and the cell surface neoantigens detected in this way are also distinctive for individual tumours.

It should be recognized that there is no formal proof that the antigens detected by *in vitro* and *in vivo* assays are identical and, although these antigens may be referred to, for simplicity, as tumour-specific antigens, it is desirable to qualify whether they were identified by immune rejection or by *in vitro* assay. This distinction is now becoming less rigid, however, since tumour-specific antigens isolated from tumour cells and in some cases characterized by *in vitro* tests have been shown to elicit a specific immune rejection response against transplanted tumour cells although immunoprotection by soluble antigens may be manifest at a much weaker level than that produced by intact cells.

Cell-mediated immune reactions to neoantigens on DAB-induced rat hepatic neoplasms were initially demonstrated using the colony inhibition technique [HELLSTRÖM, 1967] by showing that lymph node cells from rats immunized against syngeneic tumour transplants were specifically inhibitory for cells of the immunizing tumour [BALDWIN and EMBLETON, 1971b]. This technique has subsequently been modified as a microcytotoxicity assay in which the survival of tumour cells plated in wells of microtest plates and exposed to sensitized lymphoid cells is compared with that of tumour cells exposed to normal control lymphoid cells [HELLSTRÖM *et al.*, 1971]. Again, lymph node cells from rats immunized against individual DAB-induced hepatic neoplasms were cytotoxic for cells of the immunizing tumour.

Antibody responses against cell surface expressed neoantigens on DAB-induced hepatic tumours have been detected employing either the colony inhibition or microcytotoxicity assay to determine the complement-dependent cytotoxicity of serum from tumour-immune rats [BALDWIN and EMBLETON, 1971b; BALDWIN *et al.*, 1973a]. In addition, tumour-specific antibody can be detected by the membrane immunofluorescence staining of target tumour cells in suspension, and this assay has been extensively employed for characterizing cell surface expressed neoantigens on rat hepatic neoplasms [BALDWIN *et al.*, 1971; BALDWIN, 1973]. All of the DAB-induced tumours examined elicited a specific antibody response when syngeneic rats were immunized against transplanted tumour cells. These responses were highly specific so that tumour-immune serum reacted with cells of the immunizing tumour, but not with other DAB-induced hepatic tumours. Furthermore, antibody in serum from rats immunized against individual hepatic tumours could only be absorbed out with cells of the immunizing tumour and this assay has subsequently been employed in characterizing neoantigens in subcellular fractions of tumour [BALDWIN and GLAVES, 1972a; BALDWIN, 1973].

Tumour-specific antibody has also been demonstrated by membrane immunofluorescence methods in serum of rats immunized against syngeneic transplants of AAF-induced hepatic neoplasms [BALDWIN and EMBLETON, 1971a]. It should be noted that only those examples where tumour-specific rejection reactions were demonstrated showed significant antibody responses, and these were again directed specifically against cells of the immunizing tumour.

Tumour-Associated Embryonic Antigens

Re-expression of embryonic antigens on a variety of chemically induced tumours including those arising in hepatic tissues has been well documented [BALDWIN, 1973; BALDWIN et al., 1974a]. A major component arising in hepatic tumours in several species including humans, mice and rats has been identified as an α-fetoprotein [ABELEV, 1974; RUOSLAHTI et al., 1974]. This protein is found in only trace amounts in normal adult serum, but in high concentrations in sera of new-born or pregnant rats as well as rats bearing hepatic neoplasms. Elevated serum levels of α-fetoprotein have also been observed in rats during exposure to several hepatocarcinogens, often well before neoplastic change is evident. For example, BECKER and SELL [1974] demonstrated elevated α-fetoprotein levels in rats receiving approximately 1% of a carcinogenic dose of AAF.

Carcinogen-induced rat hepatic neoplasms also express embryonic antigens which unlike α-fetoprotein are immunogenic in the tumour-bearing host [BALDWIN et al., 1974a]. These embryonic antigens have been positively identified with aminoazo dye-induced rat hepatic tumours by reaction of tumour cells with either lymphoid cells or serum from multiparous female rats [BALDWIN et al., 1974a]. During pregnancy, female rats are thought to become sensitized to a wide range of antigens expressed upon cells during specific stages of embryonal development, but which are not present upon cells of the adult host [BALDWIN and VOSE, 1974b]. This can be demonstrated, for example, in tests showing positive membrane immunofluorescence staining of multiparous rat sera with cells taken from 14- to 15-day-old rat fetuses, but not with cells taken from older fetuses [BALDWIN et al., 1974a; BALDWIN and VOSE, 1974b]. In a similar manner, embryonic antigens have been detected upon aminoazo dye-induced rat hepatic tumours by membrane immunofluorescence staining or complement dependent cytotoxicity of multiparous rat serum for tumour cells. Lymph node cells from multiparous rats were also cytotoxic for the rat hepatic tumour cells [BALDWIN et al., 1974a].

Tumour-associated embryonic antigens have since been demonstrated on rat hepatic tumours induced by AAF by reaction with multiparous rat serum, irrespective of whether tumour rejection antigens are expressed [BALDWIN and VOSE, 1974a]. Comparably, cross-reacting embryonic antigens have been identified on a wide range of tumour types in the rat including carcinogen-induced or spontaneous mammary carcinomas [BALDWIN et al., 1974a; BALDWIN and EMBLETON, 1974] and carcinogen-induced colon carcinomas [STEELE and SJÖGREN, 1974]. These findings suggest that embryonic antigen expression may be a concomitant of neoplastic change, and so typing of these antigens could provide a suitable preliminary screening assay for carcinogen-induced transformation. This is exemplified by studies showing that mouse prostate cells transformed *in vitro* by chemical carcinogens generally express embryonic antigens [EMBLETON and HEIDELBERGER, 1974].

It has been possible to distinguish between the tumour-associated embryonic antigens and tumour-specific cell surface antigens by analysis of their specificities [BALDWIN, 1973; BALDWIN et al., 1974a]. Thus, as already indicated, the tumour-specific antigens are characteristic components of individual tumours. In comparison, the embryonic antigens are cross-reactive, and this has been confirmed by absorption assays on the membrane immunofluorescence staining of multiparous rat serum. These tests showed that antibody in multiparous rat serum which reacted with several tumour types could be totally absorbed with each of the reactive tumours. This indicates that reaction with multiparous serum detects common embryonic antigens expressed upon several tumour types including rat hepatic tumours as well as MCA-induced sarcomas [BALDWIN et al., 1974a]. Differentiation between tumour-specific and tumour-associated embryonic antigens expressed on DAB-induced hepatic tumours has also been possible by comparing the capacity of sera from multiparous and tumour-immune rats to block tumour cells from attack by sensitized lymph node cells [BALDWIN et al., 1974b]. These tests showed that the *in vitro* cytotoxicity of multiparous rat lymph node cells for tumour or embryo cells could be blocked by pretreating target cells with multiparous rat serum. Lymph node cells from tumour-immune rats are also cytotoxic for both tumour and embryo cells, but only the reaction with embryo cells was blocked by multiparous rat serum.

It is clear that many, if not all, chemically induced tumours express embryonic antigens detectable using reagents from multiparous donors syngeneic with the tumour. However, their role in tumour-immune reactions, and in particular, tumour rejection reactions, remains equivocal. There are contrasting reports with virus-induced tumours with regard to the efficacy of promoting a tumour rejection response by immunization with fetal tissues. For example, with SV40-induced tumours in hamsters, such procedures are effective in inducing tumour immunity [COGGIN et al., 1970, 1971], whereas with SV40 and polyoma tumours in mice, this approach is unsuccessful [TING et al., 1973]. With aminoazo dye-induced tumours in rats, attempts to induce immunity to subcutaneous challenge inocula with tumour cells have also been singularly unsuccessful [BALDWIN et al., 1974c] even though the embryo cells used for immunization (by irradiated graft implantation or excision of embryomas) were taken at 14–16 days of gestation, this being the time of maximum expression of the tumour-associated embryonic antigens on the developing embryo [BALDWIN and VOSE, 1974b]. Nevertheless, these animals were in fact sensitized to embryonic antigens following immunization, since lymph node cells from immunized donors, like those from multiparous rats, were cytotoxic *in vitro* for 15-day-old embryo cells as well as cells derived from transplanted tumours [SHAH et al., 1976]. More recently, other routes of tumour cell challenge have been employed to investigate the possibility that embryonic antigens may invoke tumour rejection reactions. Using an artificial model of metastasis, whereby challenge intravenously with MCA-induced rat sarcoma cells produces pulmonary tumour nodules, it was determined that by immunization with embryonic tissue, small but significant levels of protection could be achieved [REES et al., 1975]. In this system, these effects were amplified when 15-day-old embryo subcellular membranes were used as the immunizing material, control rats being treated with equivalent membrane preparations isolated from normal tissue, or older embryos which are deficient in the tumour-associated embryonic antigen. The significance of these findings requires further evaluation although they do emphasize that under restricted circumstances, tumour-associated embryonic antigens may function as rejection antigens.

Characteristics of Tumour-Specific Antigens

Subcellular Localization

Almost all of the *in vitro* immunological assays employed for the detection of tumour-specific antigens on rat hepatic tumours are dependent upon these determinants being expressed, at least to some extent, at the cell surface. Such tests frequently involve antibody binding to surface receptors, or cytotoxicity or cytostasis induced by lymphoid cells following interaction with target cells carrying the appropriate antigen so that, almost by definition, the targets for these reactions are exposed to the extracellular environment. However, in many cases, it is not known whether tumour-specific antigen may also be present as an intracellular moiety, possibly on a macromolecule to be inserted into the plasma membrane or an antigenically active precursor of the membrane-associated antigen. With carcinogen-induced sarcomas in the guinea-pig, for example, tumour-specific antigen was recovered in the soluble intracytoplasmic protein fraction following cell disruption so that this antigen could represent one of the cellular products just mentioned [OETTGEN et al., 1968]. Alternatively, this component may be only loosely associated with the cell surface, mechanical homogenization being sufficient to liberate the antigen as a water-soluble macromolecule. Conversely, with DAB-induced hepatomas in the rat, no evidence has been obtained to suggest that tumour-specific antigenic activity is associated with the soluble fraction of tumour homogenates, when assessed either by the capacity of these fractions to induce tumour-specific antibody formation or to inhibit the binding of specific antibody in tumour-immune serum with tumour cell surface antigen [PRICE and BALDWIN, 1974a, b]. The subcellular localization of the tumour-specific antigen in rat hepatomas would appear to be similar to that of several histocompatibility antigens studied, which show predominant expression at the cell surface but, although murine H-2 histocompatibility antigens on nuclear membranes have been described [ALBERT and DAVIES, 1973], no tumour-specific antigen activity was detected in rat hepatoma nuclei or nuclear membranes [PRICE and BALDWIN, 1974a, b].

Biochemical Characteristics

With aminoazo dye-induced rat hepatic tumours several investigations have indicated that the tumour-specific antigen is an integral membrane protein or, perhaps more probably, a membrane glycoprotein. In experiments upon subcellular fraction, antigenic activity was only demonstrated in plasma membrane containing preparations [PRICE and BALDWIN, 1974a, b], although it was possible to solubilize this activity using papain [HARRIS et al., 1973]. Separation of these soluble fractions by various chromatographic and density gradient sedimentation techniques revealed heterogeneity in molecular size of the components retaining the antigen determinants [BALDWIN et al., 1973b].

It has been possible, however, to define several parameters related to the molecular characteristics of the tumour-specific antigen isolated following enzymic digestion of cell membranes. Firstly, molecular weight estimations by sucrose density gradient centrifugation and gel filtration have indicated a value of 50,000–60,000 daltons for the smallest component displaying antigenic activity [BALDWIN et al., 1973b]. Disc gel electrophoresis has suggested that this antigen may be obtained substantially free from contaminating proteins and an isoelectric point of approximately 4.5 has been estimated. Compositional analysis has revealed amino acid profiles similar to those of rat, mouse and human histocompatibility antigens, and these data are entirely consistent with the view that this determinant is expressed upon a membrane-associated protein or glycoprotein.

Relationship to Other Cellular Antigens

Increasing attention is being paid to determine to what extent the appearance of a tumour-specific antigen on chemically induced tumours such as rat hepatic tumours is reflected in the deletion or modification of specific macromolecular products present in cells of their normal counterpart. In this respect, deletion of normal liver-specific antigens from aminoazo dye-induced rat hepatomas was demonstrated by membrane immunofluorescence techniques using appropriately absorbed rabbit anti-liver homogenate antisera [BALDWIN and GLAVES, 1972b]. It was also possible to conclude from this study that neoplastic transformation by DAB resulted in the replacement of a characteristic normal liver component by the individually distinct tumour-specific antigen. That there may be a relationship between the expression of tumour-specific antigens associated with chemically induced tumours, and normal histocompatibility antigens. HAYWOOD and McKHANN [1971] observed that a reciprocal relationship may exist between tumour-specific immunogenicity and histocompatibility H-2 antigen expression, and similar findings have been obtained using polyoma virus-induced murine tumours [TING and HERBERMAN, 1971]. More recently, it has been proposed that tumour-specific antigens arise as a result of a mutation at the level of histocompatibility genes, with the appearance of modified antigens differing in specificity but retaining the general properties of the original molecule [PARMIANI and INVERNIZZI, 1975]. This proposal was derived from experiments involving the detection of tumour-specific antigens on MCA-induced murine sarcomas which showed cross-reactivity with what were termed 'alien' histocompatibility antigens.

Another approach to determine the relationship of tumour-specific antigens to histocompatibility antigens has been to regard this question as an immunochemical problem. Purified and radioiodinated preparations of one rat hepatoma (D23)-specific antigen were found to bind specifically to a Sepharose 4B column to which syngeneic immune antiserum was coupled [BOWEN and BALDWIN, 1975]. This material, after elution by high salt, rebound to the same column and also to an immunoabsorbent column of an immobilized alloantiserum raised in another strain of rats. This was taken to reflect the fact that the molecule retaining the tumour-specific antigen determinant associated with this hepatoma may share points of identity with an histocompatibility antigen expressed upon normal rat liver.

Characteristics of Tumour-Associated Embryonic Antigens

Subcellular Localization

A biologically important site for embryonic antigen localization is at the tumour cell surface so that such determinants may function as targets for immune reactions. This has been substantiated in a series of investigations upon a variety of DAB-induced hepatomas in the rat, where it was found that in a reasonable proportion of tests, cell surface membrane immunofluorescence reactions and complement-dependent serum and lymph node cell cytotoxicity against target tumours were observed using lymphoid cells or serum from syngeneic multiparous rats [BALDWIN et al., 1974a]. With these tumours, however, embryonic antigen activity is also associated with the soluble intracytoplasmic fraction of tumour cell homogenates [BALDWIN et al., 1974a; PRICE, 1974]. The significance of this finding is as yet unknown, although it is clear that evaluation of the relationship between the soluble and membrane associated embryonic antigens of apparently the same specificity is required.

Biochemical Characteristics

Since embryonic antigens are released into the soluble fraction following cellular disruption of rat hepatic tumours, clearly this material represents a convenient source of embryonic antigens for further characterization. For example, with DAB-induced rat hepatomas, soluble tumour cell sap fractions have been subjected to sequential biochemical separative procedures so that relatively purified preparations retaining antigen activity were obtained [BALDWIN et al., 1974a]. Critical analysis of these fractions revealed that this material represented at least two distinct but co-purifying macromolecules (approximate molecular weight range, 60,000–70,000 daltons), and further fractionation is required to identify the component retaining embryonic antigen activity.

Conclusions

Carcinogen-induced rat hepatic neoplasms express a variety of neoantigens, although the most consistently demonstrable components are tumour-associated embryonic antigens. These have been demonstrated upon all of the hepatic tumours so far analyzed, including both primary and transplanted tumours induced with aminoazo dyes and AAF-induced tumours. These observations, together with the now considerable literature on embryonic antigen expression on a range of tumour types with widely differing aetiologies [BALDWIN, 1973] suggests that typing of these antigens may provide suitable methods for characterizing transformed cells. In addition, studies showing the appearance of embryonic antigens on cultured cells transformed *in vitro* by chemical carcinogens suggests that antigen assay in these systems may provide relatively simple screening systems for preliminary evaluation of chemical carcinogens.

The most significant characteristic of carcinogen-induced tumours, including those arising in hepatic tissues, is the expression of the tumour-specific cell surface antigens which may function as tumour rejection antigens. This type of neoantigen is not always expressed upon transformed cells, so that it cannot be used as a criterion of malignant change. These neoantigens are highly polymorphic and in most instances it has been shown that tumours express individually distinct components. Although biochemical characterization of these tumour cell surface products is still at an elementary level, the available evidence does suggest the possibility that they are modified histocompatibility antigens. This, together with the suggestion that the expression of this type of neoantigen requires specific interaction with carcinogen, indicates that these tumour cell products may be utilized for the analysis of 'mutation-like' events in hepatocarcinogenesis.

References

ABELEV, G.I.: α-Fetoprotein as a marker of embryo-specific differentiations in normal and tumor tissues. Transplantn Rev. *20:* 3–37 (1974).

ALBERT, W.H.W. and DAVIES, D.A.L.: H-2 antigens on nuclear membranes. Immunology 24: 841–850 (1973).

BALDWIN, R.W.: Studies on rat liver cell antigens during the early stages of azo dye carcinogenesis. Br. J. Cancer 16: 749–756 (1962).

BALDWIN, R.W.: Immunological aspects of chemical carcinogenesis. Adv. Cancer Res. 18: 1–75 (1973).

BALDWIN, R.W.: Role of immunosurveillance against chemically induced rat tumours. Transplantn Rev. 28: 62–74 (1976).

BALDWIN, R.W. and BARKER, C.R.: Tumour-specific antigenicity of aminoazo-dye induced rat hepatomas. Int. J. Cancer 2: 355–364 (1967).

BALDWIN, R.W.; BARKER, C.R.; EMBLETON, M.J.; GLAVES, D.; MOORE, M., and PIMM, M.V.: Demonstration of cell surface antigens on chemically induced tumors. Ann. N.Y. Acad. Sci. 177: 268–278 (1971).

BALDWIN, R.W. and EMBLETON, M.J.: Immunology of 2-acetylaminofluorene-induced rat mammary adenocarcinomas. Int. J. Cancer 4: 47–53 (1969).

BALDWIN, R.W. and EMBLETON, M.J.: Tumor-specific antigens 2-acetylaminofluorene-induced rat hepatomas and related tumors. Israel J. med. Scis 7: 144–153 (1971a).

BALDWIN, R.W. and EMBLETON, M.J.: Demonstration by colony inhibition methods of cellular and humoral immune reactions to tumour-specific antigens associated with aminoazo-dye-induced rat hepatomas. Int. J. Cancer 7: 17–25 (1971b).

BALDWIN, R.W. and EMBLETON, M.J.: Neoantigens on spontaneous and carcinogen-induced rat tumours defined by *in vitro* lymphocytotoxicity assays. Int. J. Cancer 13: 433–443 (1974).

BALDWIN, R.W.; EMBLETON, M.J.; PRICE, M.R., and VOSE, B.M.: Embryonic antigen expression on experimental rat tumours. Transplantn Rev. 20: 77–99 (1974a).

BALDWIN, R.W.; EMBLETON, M.J., and ROBINS, R.A.: Cellular and humoral immunity to rat hepatoma-specific antigens correlated with tumour status. Int. J. Cancer 11: 1–10 (1973a).

BALDWIN, R.W. and GLAVES, D.: Solubilization of tumour-specific antigen from plasma membrane of an aminoazo-dye-induced rat hepatoma. Clin. exp. Immunol. 11: 51–56 (1972a).

BALDWIN, R.W. and GLAVES, D.: Deletion of liver-cell surface membrane components from aminoazo-dye-induced rat hepatomas. Int. J. Cancer 9: 76–85 (1972b).

BALDWIN, R.W.; GLAVES, D., and VOSE, B.M.: Differentiation between the embryonic and tumour specific antigens on chemically-induced rat tumours. Br. J. Cancer 29: 1–10 (1974b).

BALDWIN, R.W.; GLAVES, D., and VOSE, B.M.: Immunogenicity of embryonic antigens associated with chemically-induced rat tumours. Int. J. Cancer 13: 135–142 (1974c).

BALDWIN, R.W.; HARRIS, J.R., and PRICE, M.R.: Fractionation of plasma membrane-associated tumour-specific antigen from an aminoazo-dye-induced rat hepatoma. Int. J. Cancer 11: 385–397 (1973b).

BALDWIN, R.W. and PIMM, M.V.: Proc. 7th Int. Symp. Immunopathology, Bad Schachen 1976, pp. 397–410 (Schwabe, Basel 1976).

BALDWIN, R.W. and VOSE, B.M.: Embryonic antigen expression on 2-acetylaminofluorene induced and spontaneously arising rat tumours. Br. J. Cancer 30: 209–214 (1974a).

BALDWIN, R.W. and VOSE, B.M.: The expression of a phase specific foetal antigen on rat embryo cells. Transplantation 18: 525–530 (1974b).

Bartlett, G.L.: Effect of host immunity on the antigenic strength of primary tumors. J. natn. Cancer Inst. *49:* 493–504 (1972).

Becker, F.F. and Sell, S.: Early elevation of α_1-fetoprotein in N-2-fluorenylacetamide hepatocarcinogenesis. Cancer Res. *34:* 2489–2494 (1974).

Bowen, J.G. and Baldwin, R.W.: Tumour-specific antigen related to rat histocompatibility antigens. Nature, Lond. *258:* 75–76 (1975).

Coggin, J.H.; Ambrose, K.R., and Anderson, N.G.: Fetal antigen capable of inducing transplantation immunity against SV40 hamster tumor cells. J. Immun. *105:* 524–526 (1970).

Coggin, J.H.; Ambrose, K.R.; Bellomy, B.B., and Anderson, N.G.: Tumor immunity in hamsters immunized with fetal tissues. J. Immun. *107:* 526–533 (1971).

Embleton, M.J. and Heidelberger, C.: Antigenicity of clones of mouse prostate cells transformed *in vitro*. Int. J. Cancer *9:* 8–18 (1972).

Embleton, M.J. and Heidelberger, C.: Neoantigens on chemically transformed cloned C3H mouse embryo cells. Cancer Res. *35:* 2049–2055 (1974).

Gordon, J.: Isoantigenicity of liver tumours induced by an azo dye. Br. J. Cancer *19:* 387–391 (1965).

Harris, J.R.; Price, M.R., and Baldwin, R.W.: The purification of membrane-associated tumour antigens by preparative polyacrylamide gel electrophoresis. Biochim. biophys. Acta *311:* 600 (1973).

Haywood, G.R. and McKhann, C.F.: Antigen specificities on murine sarcoma cells. Reciprocal relationship between normal transplantation antigens (H-2) and tumor-specific immunogenicity. J. exp. Med. *133:* 1171–1187 (1971).

Hellström, I.: A colony inhibition (CI) technique for demonstration of tumor cell destruction by lymphoid cells *in vitro*. Int. J. Cancer *2:* 65–68 (1967).

Hellström, I.; Hellström, K.E.; Sjögren, H.O., and Warner, G.A.: Demonstration of cell-mediated immunity to human neoplasms of various histological types. Int. J. Cancer *7:* 1–16 (1971).

Ishidate, M.: Antigenic specificity of hepatoma cell lines derived from a single rat. Abstr. Int. Cancer Congr., Houston, 1970, p. 227.

Kitagawa, M.; Yagi, Y.; Planinsek, J., and Pressman, D.: *In vivo* localization of anticarcinogen antibody in organs of carcinogen-treated rats. Cancer Res. *26:* 221–227 (1966).

Mondal, S.; Iype, P.T.; Griesbach, L.M., and Heidelberger, C.: Antigenicity of cells derived from mouse prostate cells after malignant transformation *in vitro* by carcinogenic hydrocarbons. Cancer Res. *30:* 1593–1597 (1970).

Oettgen, H.F.; Old, L.J.; McLean, E.P., and Carswell, E.A.: Delayed hypersensitivity and transplantation immunity elicited by soluble antigens on chemically induced tumours in inbred guinea pigs. Nature, Lond. *220:* 295–297 (1968).

Parmiani, G. and Invernizzi, G.: Alien histocompatibility determinants on the cell surface of sarcomas induced by methylcholanthrene. I. *In vivo* studies. Int. J. Cancer *16:* 756–767 (1975).

Prehn, R.T.: Relationship of tumor immunogenicity to concentration of oncogen. J. natn. Cancer Inst. *55:* 189–190 (1975).

Price, M.R.: Isolation of embryonic antigens from chemically induced rat hepatomas. Biochem. Soc. Trans. *2:* 650–652 (1974).

Price, M.R. and Baldwin, R.W.: Preparation of aminoazo dye-induced rat hepatoma membrane fractions retaining tumour specific antigen. Br. J. Cancer *30:* 382–393 (1974a).

PRICE, M.R. and BALDWIN, R.W.: Immunogenic properties of rat hepatoma subcellular fractions. Br. J. Cancer *30:* 394–400 (1974b).

PRICE, M.R. and BALDWIN, R.W.: Shedding of tumour cell surface antigens. Cell Surface Rev. *6:* 177–207 (1977).

REES, R.C.; SHAH, L.P., and BALDWIN, R.W.: Inhibition of pulmonary tumour development in rats sensitized to rat embryonic tissue. Nature Lond. *255:* 329–330 (1975).

RUOSLAHTI, E.; PIHKO, H., and SEPPALA, M.: Immunochemical purification and chemical properties. Expression in normal state and in malignant and non-malignant liver disease. Transplantn Rev. *20:* 38–60 (1974).

SHAH, L.P.; REES, R.C., and BALDWIN, R.W.: Tumour rejection responses in rats sensitized to rat embryonic tissue. I. Rejection of tumour cells implanted subcutaneously and detection of cytotoxic lymphoid cells in embryo-sensitized rats. Br. J. Cancer *33:* 577–583 (1976).

STEELE, G. and SJÖGREN, H.O.: Embryonic antigens associated with chemically induced colon carcinomas in rats. Int. J. Cancer *14:* 435–444 (1974).

STUTMAN, O.: Immunodepression and malignancy. Adv. Cancer Res. *22:* 261–422 (1975).

TING, C.-C. and HERBERMAN, R.B.: Inverse relationship of polyoma tumour specific cell surface antigen to H-2 histocompatibility antigens. Nature new Biol. *232:* 118–120 (1971).

TING, C.-C.; RODRIGUES, D., and HERBERMAN, R.B.: Expression of fetal antigens and tumor-specific antigens in SV40-transformed cells. II. Tumor transplantation studies. Int. J. Cancer *12:* 519–523 (1973).

Prof. R.W. BALDWIN, Cancer Research Laboratories, University of Nottingham, *Nottingham* (England)

Subversion of the Immune System by Tumors as a Mechanism of their Escape from Immune Rejection

Otto J. Plescia, Kazimiera Grinwich, John Sheridan and Anne M. Plescia

Waksman Institute of Microbiology, Rutgers, The State University of New Jersey, New Brunswick, N.J.

Introduction

The notion that autochthonous tumors possess tumor-specific, or at least tumor-associated, antigens sufficiently different from normal tissue antigens to render these tumors susceptible to immunological attack by the host is hardly new. The idea goes back at least to the time when it had been established that infectious diseases could be controlled by immunological means.

Ehrlich, for example, intuitively regarded an autochthonous tumor as a foreign antigenic mass subject to attack and rejection by its host, much like an infectious bacterium. He addressed himself to this question of the antigenicity of tumors at a meeting of the British Royal Society in 1900. He acknowledged the lack of any decisive data to support his position, but he argued that 'even if in the immediate future no great practical success is attained, we must remember that we are only at the beginning of a rational investigation of properties of (tumor) cells which hitherto have been far too lightly regarded'.

Despite this kind of exhortation by Ehrlich, progress in tumor immunology was understandably slow. The task of establishing tumor-associated transplantation antigens – a necessary first step in a rational approach to immunological control of cancer – was fraught with pitfalls until inbred strains of experimental animals became generally available and the laws of transplantation biology were established.

Some 20 years have elapsed since the first decisive reports of tumor-specific transplantation antigens [1]. Yet the goal of immunological control of cancer remains on the horizon, an elusive prize to be won. Since these

Table I. Possible reasons for failure of immune surveillance system against cancer

I. Properties of tumor antigens
   A. Antigens are weak and poorly immunogenic
   B. Antigens are sheltered from immunocompetent cells of the host because of the location of the tumor
   C. Antigens induce specific unresponsiveness depending on the concentration and distribution of the antigens
   D. Antigens induce specific blocking factors that interfere with tumor immunity
II. Host competence
   A. Immunodeficiency prior to tumor formation
      1. Genetic
      2. Environmental
   B. Immunodeficiency associated with tumorigenesis, i.e. tumor cells are generally immunosuppressive

first reports, we have been able to identify and characterize the different types of cells that comprise the immune system, including humoral and cellular immunity, and the non-specific phagocytic system. We have come to understand the mechanism of homograft rejection and, assuming a tumor may be regarded as a homograft, one would also expect to know the mechanism of tumor rejection.

Ironically, the rejection of a normal homograft is so efficient that heroic measures need to be taken to prevent rejection, but the rejection of an autochthonous or syngeneic tumor is so inefficient that heroic means are being sought to stimulate the rejection of these tumors. The assumption, therefore, that a tumor is essentially a homograft cannot be entirely valid. A tumor, like a homograft, may possess transplantation antigens, but the similarity stops there. A tumor, unlike a homograft of normal tissue, seems able to escape from immune rejection. The question is why and how do tumors escape? It is a key question because further progress in the immunotherapy of cancer is unlikely without an answer to this question.

There are several possible reasons why an autochthonous or syngeneic tumor fails to elicit an effective immune response or why the host is unable to respond effectively. These are listed in table I. One of these possibilities, which is the subject of this report, is that tumors are subversive, thus inactivating their host's immunological system and creating a favorable environment for themselves in an otherwise hostile environment. There is circumstantial and direct evidence in support of this view. Cancer patients tend to become immunologically anergic as the disease progresses [2–6], as do experimental animals bearing syngeneic tumors [7, 8, 16–20]. In the latter case, the evidence is clear-cut because the experimental animals, unlike the cancer patients, were not treated with immunosuppressive anti-tumor drugs.

Heretofore, tumors have been regarded as passive with respect to the immune system, interacting with immune cells only through their antigens. There is now evidence that the opposite is true, that tumors have the biochemical capability of acting aggressively against certain types of immune cells and subverting them, using this mechanism as a means of escape from immunological rejection.

Results

Effect of Syngeneic Tumors on the Immune System of Mice

This study was designed to assess the suppressive effects of syngeneic tumors, the specific objective being to identify the types of immune cells affected. To this end several inbred strains of mice (C57B1/6J, BALB/c, DBA/2J), bearing syngeneic tumors of different etiology (chemical and viral), were challenged with sheep red blood cells (SRBC) to study antibody response, with a tumor allograft to study cellular and antibody response, with T- and B-cell mitogens to delineate further the cell types involved in cellular and antibody responses, and with viable infectious microorganisms to examine mononuclear phagocytic cells.

Antibody response of tumor-bearing mice to SRBC. Groups of C57B1/6J mice bearing the transplanted MC-16 tumor (induced in C57B1/6J mice by methylcholanthrene) and BALB/c mice bearing the MCDV-12 ascites tumor (induced in BALB/c mice by Rauscher leukemia virus) were immunized with SRBC. A single injection of SRBC was administered intraperitoneally after implantation of the syngeneic tumor, the time ranging to 20 days. Groups of normal mice without tumor were similarly immunized, and they served as controls. There was evidence of depression of the antibody response to SRBC in the tumor-bearing mice by the fifth day, and by the 10th day the antibody response was essentially absent [7]. Of course, by this time the tumors had become well established and were growing progressively in size. Since the tumors and SRBC were antigenically unrelated, it was concluded that generalized immunodepression is at least associated with, if not the result of, growth of a syngeneic tumor. These results also suggested that this immunodepression is common to syngeneic tumors regardless of etiology; in this study, both a chemical and a virus-induced tumor line were used.

It was important to establish the role, if any, of the syngeneic tumor in the suppression of the antibody response of the tumor-bearing mice to SRBC. To this end, 10^7 normal C57Bl/6J spleen cells were cultured *in vitro* with MC-16 syngeneic tumor cells, ranging in number from 10^3 to 10^5, and SRBC were added directly to these cultures to induce antibody formation, measured in terms of the development of hemolytic plaque-forming cells 4 days after the start of the cultures. The tumor cells proved very active in suppressing antibody response; 10^5 tumor cells completely suppressed this response, and the addition of as few as 1 tumor cell per 1,000 spleen cells resulted in significant suppression [1]. This was a simple defined system consisting only of lymphoid spleen cells and tumor cells. Thus, the results of this kind of experiment provided clear, direct evidence of the immunosuppressive activity of syngeneic tumor cells.

In the above experiment, viable tumor cells, prepared from freshly excised tumors of passage mice, were used, and they were added to cultures of spleen cells essentially at the same time as SRBC. This kind of experiment was repeated, except that non-viable mitomycin C-treated tumor cells were also included and the addition of SRBC to the culture was delayed up to 24 h to allow tumor and spleen cells to interact with one another. The extent of immunosuppression by the tumor cells increased with increased time of interaction, and it decreased with loss of viability [7]. These results provide further evidence of the direct immunosuppressive activity of syngeneic tumor cells, and they also indicate the dynamic nature of this immunosuppressive activity.

Cellular immune response of tumor-bearing mice. The finding of a generalized depression of the antibody response in tumor-bearing mice was considered to be, not a fortuitous event, but a reflection of a mechanism that tumors might use to escape from immunological rejection by subverting the immune system. This hypothesis prompted a study of the effect of a syngeneic tumor on the development of cellular immunity since specific tumor immunity is largely dependent on cell-mediated immunity rather than on antibody-mediated immunity.

For this purpose, the cellular immune response to a tumor allograft was studied in a histoincompatible strain of mice bearing a transplanted syngeneic tumor. Tumor cells of the MC-16 line were inoculated subcutaneously in syngeneic C57Bl/6J mice, and at different times thereafter groups

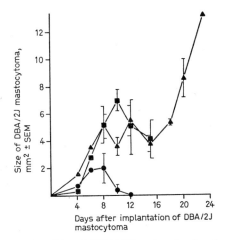

Fig. 1. Failure of C57Bl/6J strain mice, bearing a syngeneic tumor, to reject a tumor allograft. Syngeneic tumor cells (MC-16) were implanted subcutaneously into two groups of C57Bl/6J mice; a control group received diluent. All groups of mice were challenged i.d. with viable allogeneic tumor cells of the DBA/2J strain, one experimental group 5 days (▲) and the other 10 days (■) after implantation of syngeneic tumor cells, and the control group (●) 10 days after diluent was given. Growth of the DBA/2J allogeneic tumor was monitored by measuring the diameter of the tumor mass at the site of inoculation.

of these mice, and a control group that received diluent, were inoculated in a contralateral site with allogeneic DBA/2J mastocytoma cells.

The objective of this type of experiment was to assess the effect of a growing viable syngeneic tumor on the capacity of its host to reject a tumor allograft. From the growth profile of the DBA/2J tumor allograft in C57Bl/6J mice as a function of the interval of time between inoculation of the syngeneic tumor and challenge with tumor allograft, it was clear that the tumor allograft grew faster and more extensively in these mice, and more important it was not rejected by mice bearing a syngeneic tumor for only 5 days at the time the tumor allograft was inoculated (figure 1). These results may be summarized by saying that a syngeneic tumor can so depress the cellular immune capability of its host that dividing tumor cells, even strongly antigenic allogeneic tumor cells, can and do escape immunological rejection.

Identification of target lymphoid cells. Specific rejection of a homograft, including a tumor allograft, is known to be mediated by T cells. In contrast, the antibody response to SRBC requires the active participation of both T and B cells. Thus, the failure of mice bearing a syngeneic tumor to reject a tumor allograft and to develop a normal antibody response to SRBC clearly implicates T cells as targets of immunosuppressive syngeneic tumor cells. Regarding the B cells, these results are indecisive because failure of the antibody response to SRBC would occur whether or not the B cells are inactivated.

To explore further the identity of the target lymphoid cells, mitogens that stimulate blastogenesis of either T or B cells preferentially, i.e. PHA-M (phytohemagglutinin) and LPS (*E. coli* lipopolysaccharide, T and B cell mitogens, respectively, were used as probes.

Groups of C57B1/6J mice were inoculated subcutaneously with syngeneic MC-16 tumor cells, and control groups received diluent. At times thereafter, up to 20 days, several mice from both the control and experimental groups were sacrificed, their spleens were excised, and pooled suspensions of spleen cells were prepared from them to be cultured *in vitro*. To replicate cultures of these spleen cells (10^7 culture) was added either PHA-M, LPS, or diluent. After 20 h of culture. ^{3}H-thymidine was added, and the cultures were continued in culture for an additional period of 20 h during which ^{3}H-thymidine was incorporated into newly synthesized DNA resulting from mitogenic stimulation. Thus, net incorporation of ^{3}H-radioactivity into DNA provides a measure of the blastogenic response of T and B cells to appropriate mitogenic stimulation.

The results showed a differential effect of the syngeneic tumor on T and B cells [8]. During tumor growth, there was a progressive loss of responsiveness of spleen cells to stimulation by PHA-M, but no significant change in their response to LPS. In a continuation of this study, a variable number of viable tumor cells were added directly to *in vitro* cultures of normal syngeneic tumor cells (10^7/culture). Tumor and spleen cells were allowed to interact for several minutes, after which mitogen (PHA-M or LPS) was added, and stimulation by mitogen was assessed as before in terms of incorporation of ^{3}H-thymidine into DNA. Necessary controls, including cultures of spleen cells or tumor cells alone and cultures of cells without mitogen, were included. The results of these additional experiments confirmed the preferential subversive activity of a syngeneic tumor against T cells observed *in vivo* [8]. This study was extended to include other syngeneic tumor cell lines of chemical and viral etiology. In each case, a fixed number of tumor cells (10^5) were added to cultures of 10^7 normal syngeneic spleen cells, and the time of addition of the mitogen after the start of the culture was varied. The results were graphed as the percent suppression of mitogenic response as a function of the time of interaction between tumor and

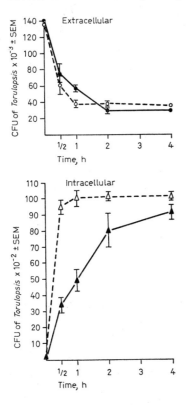

Fig. 2. Effect of a syngeneic tumor on the phagocytic activity *in vitro* of spleen cells from BALB/c mice. Spleen cells (1×10^7)/culture) from normal mice and mice bearing the MCDV-12 tumor for 10 days were infected with 10^5 *Torulopsis* and incubated at 37°C for phagocytosis to occur. At times indicated the cultures were examined for intracellular (glass adhering, phagocytized) *Torulopsis*, and for extracellular (non-adhering) *Torulopsis*, in terms of colony forming units (CFU). Open symbols, O and △, are for spleen cells from tumor-bearing mice, and closed symbols, ● and ▲, are for normal spleen cells.

spleen cells before addition of mitogen, and from this graph could be interpolated the time of interaction required for the mitogenic response to be reduced to 50% of the control level. This time should decrease with increase in the immunosuppressive activity of the tumor and therefore is an index of the immunosuppressive nature of a tumor line. Indeed, all the tumor lines tested showed a capacity to subvert T cells, but they differed quantitatively in this respect.

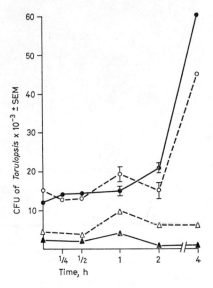

Fig. 3. The phagocytic activity of normal spleen cells after interaction with syngeneic tumor cells *in vitro*. Normal BALB/c spleen cells (10^7) were cultured with MCDV-12 tumor cells (10^5) for 24 h at 37°C, after which 10^6 *Torulopsis* were added and the cultures were incubated for 1 h for phagocytosis to occur. Phagocytosis was stopped by washing away excess nonphagocytized *Torulopsis*, and the number of *Torulopsis* phagocytized by adherent cells was determined as colony-forming units (CFU). Cultures containing only spleen cells and tumor cells were included as controls, as were cultures with *Torulopsis* alone. Key to the symbols: O = spleen + tumor cells; ● = spleen cells; ▲ = tumor cells; △ = *Torulopsis*.

Phagocytic cells as targets of syngeneic tumor cells. The mononuclear phagocytic cells constitute an important first line of defense against microbial infection, and particulate antigens in general. There is increasing evidence that these cells may even afford protection against syngeneic and autochthonous tumors. As you might expect, therefore, tumor-bearing mice (several strains and different tumor lines) were tested for changes in resistance to microbial infection, and phagocytic cells of the spleen and peritoneal exudate from tumor-bearing animals were examined for their ability to phagocytize and kill test microorganisms in *in vitro* culture. The direct action of tumor cells was also examined, adding them to *in vitro* cultures of spleen cells. The same conclusion could be drawn from the results of all these experiments; the phagocytic cells were not depressed by the action of the syngeneic tumors, and if they were changed at all, they seemed somewhat more active phagocytically (fig. 2, 3).

This finding was surprising in view of reports of tumor-derived factors that repulse macrophages [9], and the apparent inactivity of macrophages found in tumor masses [10]. Nevertheless, the fact remains that syngeneic tumors do not subvert phagocytic cells under the same conditions and in the same test systems in which they subvert T cells.

Mechanism of Subversion of T Cells by Syngeneic Tumors

Role of prostaglandins as mediators. Prostaglandins were first considered and tested as possible mediators of immunosuppression by syngeneic tumors because several tumor lines have been reported to synthesize and release abnormal amounts of prostaglandins [11–13], which by themselves are reported to influence immunological responses [14].

Indeed, the addition of prostaglandins of the E series (PGE_1 and PGE_2) to cultures of normal spleen cells mimicked syngeneic tumor cells in suppressing the antibody response of the spleen cells to SRBC [7]. More importantly, the addition of indomethacin, an inhibitor of prostaglandin synthetases, to cultures of tumor and spleen cells reduced significantly the suppression of the antibody response to SRBC. These results were regarded as an important first clue to the role of prostaglandins in the subversion of the immune system by syngeneic tumors, and this study was extended to include the effect of prostaglandins on the mitogenic response of normal spleen cells and also the effect of inhibitors of prostaglandin synthetases on tumor-mediated suppression of the mitogenic response.

To cultures of spleen cells were added either PGE_2 or PGE_{2a}, in different amounts. 5 min later, mitogen (PHA-M or LPS) was added, and mitogenic stimulation was assessed as before. Only PGE_2, which is the series of prostaglandins produced in greatest amount by tumor cells, showed a suppressive effect, and like tumor cells it suppressed preferentially the T cell response to PHA-M [8]. There was, however, a quantitative difference between PGE_2 and tumor; at the highest non-toxic concentration of PGE_2 complete suppression of the mitogenic response was not achieved, in contrast to the complete suppression obtained with 10^5 tumor cells. This quantitative difference might be technical in nature, due to the difficulty in simulating the distribution and concentration of PGE_2 that develops in the microenvironment of interacting tumor and spleen cells. Or indeed PGE_2 is only one of the many products in the prostaglandin pathway that are capable of subverting T cells. Also of importance might be the instability of some of the prostaglandins; the thromboxanes, for example, have an extremely short half-life. Therefore, this problem of assessing prostaglandins was also approached through the use of inhibitors of prostaglandin synthetases. Three such inhibitors were used; namely indomethacin, aspirin, and flufenamic acid. These were added to cultures of tumor and spleen cells at the start, and the mitogen PHA-M was added 4 h later. All inhibited significantly the tumor-mediated suppression of the mitogenic response of the spleen cells. Because they all share in common the property of inhibiting prostaglandin synthesis, one can reasonably assign some role to the prostaglandins as mediators of immunosuppression by tumor cells. This is further supported by the finding that the extracellular concentration of PGE increased sharply with time in cultures of tumor and spleen cells and that indomethacin did indeed block the synthesis of PGE in equivalent cultures. Moreover, PGE did not suppress the phagocytic activity of adherent spleen cells, and neither did syngeneic tumor cells. Thus, in every instance thus far PGE has mimicked the immunosuppressive activity of syngeneic tumor cells.

Role of cyclic AMP as a messenger. Given the evidence that certain prostaglandins may be mediators of immunosuppression by tumor cells, the question is how do they function? The simplest explanation would be to assume a direct binding of prostaglandins by receptors on the target T cells. Lymphoid cells are known to have receptors for prostaglandins, and it is also known that such binding results in an increase in the cellular level of endogenous cAMP which can have profound effects on the activity and function of these cells. As in other instances of hormonal action, the cAMP would act as a messenger in carrying out the primary action of the prostaglandins.

This hypothesis was put to an experimental test by measuring changes in the concentration of cAMP in cultures of spleen cells to which had been added either tumor cells, PGE, agents known to stimulate the synthesis of cAMP in lymphoid cells (theophylline and isoproterenol), or exogenous dibutyrl cAMP. Large increases in the level of cAMP were noted in the response of spleen cells to PGE, theophylline and isoproterenol; little or no change was seen in cultures of tumor and spleen cells. Equivalent cultures were also assayed for the response of the spleen cells to mitogenic stimulation by PHA-M. As expected, the spleen cells cultured with tumor cells and PGE were suppressed in their mitogenic response. The addition of theophylline and dibutyrl cAMP also resulted in suppression of mitogenic stimulation, but the addition of isoproterenol was without effect. Thus, three of the four suppressive agents also stimulated the synthesis of cAMP, and one of the four cAMP-stimulating agents was not immunosuppressive. This lack of perfect correlation between stimulation of cAMP synthesis and suppression of mitogenic response is not surprising because of the gross heterogeneity of the spleen cells used and the relatively low concentration of T cells in the spleen that respond to PHA-M. One cannot be certain, for example, which cells are stimulated by isoproterenol to contribute to the bulk change in the level of cAMP and whether they include T cells that respond to mitogen. If they do not, understandably one would observe an increase in total cAMP without any suppressive effect, as was indeed the case. In cultures containing tumor cells, the problem with heterogeneity is further complicated by the presence of tumor cells that also contribute to the pool of cAMP. Nevertheless, it is noteworthy that PGE and tumor cells differ quantitatively once again, this time in their stimulation of cAMP synthesis by spleen cells. PGE stimulates cAMP synthesis and tumor cells do not. This difference may mean either that PGE is not the only mediator of immunosuppression by tumor cells, as noted already, or that tumor-mediated immunosuppression does not use the cAMP pathway exclusively, or that perhaps both deductions have merit. The evidence suggesting a role for cAMP is meager and indirect for the moment, with the exception of the finding of a significant increase in the cAMP level in the spleens of tumor-bearing mice within 2–3 days after implantation of the tumor. Another possibility is that the apparent difference between tumor cells and PGE might be technical in nature, in which case both the role of prostaglandins as mediators of immunosuppression by tumor cells and the role of cAMP as a messenger of prostaglandin activity can be accepted.

Discussion

By now there is abundant evidence, from clinical and experimental studies, that generalized immunodepression is associated with advanced stages of cancer [2–5] and that the prognosis of cancer recurring in patients following its removal by surgery correlates well with the immunological status of the patient [6]. A critical question is when does immunodepression set in relative to the period of tumorigenesis? If it develops relatively late when the tumor is fully established and growing progressively, quite clearly tumorigenesis cannot be due to a failure in the immunological system of its host. On the other hand, if immunodepression is first localized in the microenvironment of the tumor, before it manifests itself as a generalized immunodepression, tumorigenesis could be due to a localized failure of the immune system. In this latter case, immunodepression would have to be caused by the tumor itself.

Our studies, in which normal mouse spleen cells were cultured *in vitro* with syngeneic tumor cells and tested immunologically, have provided direct evidence of the immunosuppressive activity of syngeneic murine tumors. As few as 1 tumor cell added to 1,000 spleen cells suppressed the antibody response of the spleen cells to SRBC and their mitogenic response to a T cell mitogen. In a sense, this artificial mixture of tumor and spleen cells *in vitro* represents the microenvironment of the tumor *in vivo*, in which host immunocytes undoubtedly interact with tumor cells much as in the *in vitro* test system. If the spleen cells are suppressed by tumor cells *in vitro*, it is not unreasonable to expect that localized immunosuppression by a tumor also occurs *in vivo*. It would seem, therefore, that the generalized immunodepression seen in tumor-bearing animals is caused by the direct action of the tumor, that it starts as a localized immunosuppression in the microenvironment of the tumor where host immunocytes are attracted by the antigenic tumor, and that this immunosuppression permits the tumor to escape from immunological surveillance and rejection.

The fact that mice bearing a growing syngeneic tumor were unable to reject a histoincompatible tumor graft provides evidence that tumor-mediated immunosuppression can indeed provide a tumor an avenue of escape. Thus, an antigenic tumor can adapt to an immunological environment by the expedient of subverting it, and this is the reason for referring to tumor-mediated immunosuppression as subversion.

Several tumor cell lines were tested for suvbersive activity. These differed in their etiology, some being chemically induced and others virally induced. They all were subversive, although some more so than others, suggesting that this subversive property may be generally typical of tumor cells, regardless of etiology, and may be an important element in their malignancy.

In order to deal with this problem of subversion by tumors, it is essential to know which types of immunocytes are targets of subversive tumor cells and what is the mechanism of subversion.

The evidence clearly points to T cells as targets. Syngeneic tumors depress the antibody response to T-dependent antigens, the development of T cell-mediated immunity to a homograft, and the mitogenic response of spleen cells to a T cell mitogen. The B cells are presumably spared because the mitogenic response of spleen cells to a B cell mitogen was not depressed by the same tumors that depressed the T cell mitogenic response. Regarding macrophages and mononuclear phagocytes in general, there is some question.

On the one hand, in our studies tumor-bearing mice were no more susceptible to infection by microorganisms cleared by phagocytosis than normal mice, and tumor cells added to spleen cells in culture did not suppress the phagocytic and microbicidal activity of the spleen cells. On the other hand, there is evidence of a tumor-derived factor that is capable of repulsing macrophages [9], and also that macrophages infiltrate tumors without evidence of anti-tumor activity [10].

The possibility that prostaglandins might be mediators derived from and used by tumors to subvert the immune system was first suggested by published reports that tumors, both experimental and clinical, tend to synthesize inordinate amounts of prostaglandins of the E series [11–13], known to influence immunological function [14]. When tested, these prostaglandins (PGE_2) indeed mimicked the suppressive activity of tumor cells. They suppressed the antibody response of spleen cells to SRBC, and they differentially suppressed the mitogenic response of spleen cells to a T cell mitogen. Also, like tumor cells PGE_2 did not inactivate phagocytic cells in the spleen. This suggestive evidence of the role of prostaglandins as mediators of subversion by tumor cells was reinforced by the finding that our test tumor cells did in fact produce prostaglandins and that inhibitors of prostaglandin synthetases, such as indomethacin, aspirin, and flufenamic acid, blocked the subversive activity of tumor cells in *in vitro* test systems and also their synthesis of prostaglandins. However, efforts to isolate tumor-derived subversive factors from cultures of tumor cells were unsuccessful, possibly because of the instability of such expected factors or their low concentration. Conceivably, these factors might be much more efficient suppressive agents when they are released by tumor cells nearby target cells at the time of interaction. No attempt was made either to increase the number of tumor cells in culture or to concentrate the tissue culture fluid. Thus, implication of prostaglandins in tumor-mediated immunosuppression is based primarily on indirect evidence, but nevertheless a substantial body of such evidence.

Without actually isolating and identifying the responsible factors, it is impossible to know which of the many products of the prostaglandin pathways might be involved and also to exclude other classes of tumor-derived substances. Regarding this latter point, it might be significant that inhibitors of prostaglandin synthetases did not block completely the subversive activity of tumor cells, but of course it was not established that inhibition of prostaglandin synthesis was complete either. Whether or not other tumor substances contribute to tumor-mediated immunosuppression, there is little question that prostaglandins can play a significant role.

Whatever the nature of the mediators, the question is what is their mechanism of action? The fact that PGE and other chemically diverse substances, whose only common property is their ability to activate adenylate cyclase, tend to be immunosuppressive and mimic tumor cells suggests a role for cAMP as a messenger of tumor-derived mediators. A serious limitation of this hypothesis, however, is the failure to observe any increase in the concentration of cAMP in cultures of spleen and tumor cells, under conditions that result in subversion of the spleen cells [8]. This failure may, however, be technical in nature, and in fact the splenic level of cAMP increases in mice following implantation of a syngeneic tumor [15].

We, and others also [16–20], have established the general subversive nature of tumor cells and the fact that this subversion of the immune system by the tumors may provide them an avenue of escape. Clearly, to prevent this escape it is necessary to do two things: (1) to block further subversion of the immune system, and (2) to restore the depressed immunological system to full competence, and, if possible, to activate it above its normal level with non-specific immunostimulating drugs.

The major emphasis today by tumor immunologists is on a search for drugs that can activate a normal immunological system, neglecting the fact that cancer patients tend to be immunodepressed at the time of treatment. From our studies and those of others, it is evident that T cells are the targets of subversive tumor cells. Whether or not these tumor-suppressed T cells can be rescued by drug treatment remains to be explored. We have made no attempts as yet, but one possibility would be to use drugs that inhibit adenylate cyclase and/or activate cAMP-phosphodiesterase, assuming that the evidence of a role for cAMP in tumor-mediated immunosuppression is valid.

As for drugs that can block tumor-mediated immunosuppression, we already have evidence that indomethacin and aspirin, both inhibitors of prostaglandin synthetases, are effective *in vivo* in retarding the growth of a transplanted chemically induced fibrosarcoma [15] and in preventing the development of Moloney virus-induced sarcoma in mice infected with the virus [21]. These drugs seem to be effective provided that the host is immunologically competent at the time treatment is started and that the tumor is antigenic.

These are promising early results in an approach to cancer immunotherapy based on the recognition that antigenic tumor cells are not just passive sources of antigens and waiting to be rejected by their host, but are instead aggressively adapting to the hostile environment of their host by subverting the immunological system of the host. What is extraordinary is that tumor cells seem to use normal tissue hormones, in this case the prostaglandins, to achieve their mission of survival.

Acknowledgement

The grant support of the Ruth Estrin Goldberg Memorial Foundation, and a Johanna and Charles Busch Predoctoral Fellowship to K.G., are gratefully acknowledged.

References

1 PREHN, R.T. and MAIN, J.M.: Immunity to metyhlcholanthrene-induced sarcomas. J. natn. Cancer Inst. *18:* 769–778 (1957).
2 KERSEY, J.N.; SPECTOR, B.D., and GOOD, R.A.: Immunodeficiency and cancer. Adv. Cancer Res. *18:* 211–230 (1973).
3 LEHANE, D.E. and LANE, M.: Immunocompetence in advanced cancer patients prior chemotherapy. Oncology *30:* 458–466 (1974).
4 HERSH, E.M., *et al.:* Immunocompetence, immunodeficiency and prognosis in cancer. Ann. N.Y. Acad. Sci. *276:* 386–406 (1976).
5 PINSKY, C.M.; WANEBO, H.; MIKE, V., and OETTGEN, H.: Delayed cutaneous hypersensitivity reactions and prognosis in patients with cancer. Ann. N.Y. Acad. Sci. *276:* 407–410 (1976).
6 SOUTHAM, C.M.: Relationship between immunology and clinical oncology. Am. J. clin. Path. *62:* 224–242 (1974).
7 PLESCIA, O.J.; SMITH, A.H., and GRINWICH, K.: Subversion of immune system by tumor cells and role of prostaglandins. Proc. natn. Acad. Sci. USA *72:* 1848–1852 (1975).
8 PLESCIA, O.J.; GRINWICH, K., and PLESCIA, A.M.: Subversive activity of syngeneic tumor cells as an escape mechanism from immune surveillance and the role of prostaglandins. Ann. N.Y. Acad. Sci. *276:* 455–465 (1976).
9 FAUVE, R.M., *et al.:* Antiinflammatory effects of murine malignant cells. Proc. natn. Acad. Sci. USA *71:* 4052–4056 (1974).
10 EVANS, R.: Macrophages in syngeneic animal tumors. Transplantation *14:* 468–473 (1972).
11 HUMES, J. and STRAUSSER, H.: Prostaglandins and cyclic nucleotides in Moloney sarcoma tumors. Prostaglandins *5:* 183–196 (1974).
12 SYKES, J. and MADDOX, J.: Prostaglandin production by experimental tumors and effects of antiinflammatory compounds. Nature new Biol. *237:* 59–61 (1972).
13 TASHJIAN, A., jr.; VOELKEL, E.; LEVINE, L., and GOLDHABER, P.: Evidence that the bone resorption-stimulating factor produced by mouse fibrosarcoma cells is prostaglandin E_2. J. exp. Med. *136:* 1329–1343 (1972).
14 MELMON, K., *et al.:* Hemolytic plaque formation by leukocytes *in vitro:* controlled by vasoactive amines. J. clin. Invest. *53:* 13–21 (1974).

15 PLESCIA, O.J.; SMITH, A.H.; GRINWICH, K., and FEIT, C.: The problem of cancer immunotherapy in perspective. Fundamental aspects of neoplasia, pp. 139–151 (Springer, Berlin 1975).
16 WONG, A.; MANKOVITZ, R., and KENNEDY, J.C.: Immunosuppressive and immunostimulatory factors produced by malignant cells *in vitro*. Int. J. Cancer *13:* 530–542 (1975).
17 GILLETTE, R.W. and BOONE, C.W.: Changes in the mitogen response of lymphoid cells in progressive tumor growth. Cancer Res. *35:* 3774–3779 (1975).
18 SPECTER, S.; PATEL, C., and FRIEDMAN, H.: Immunosuppression induced *in vitro* by cell-free extracts of Friend leukemia virus-infected splenocytes. J. natn. Cancer Inst. *56:* 143–147 (1976).
19 KIRCHNER, H., *et al.:* Inhibition of proliferation of lymphoma cells and T lymphocytes by suppressor cells from spleens of tumor-bearing mice. J. Immun. *114:* 206–210 (1975).
20 BURK, M.; YU, S.; RISTOW, S., and MCKHANN, C.: Refactoriness of lymph-node cells from tumour-bearing animals. Int. J. Cancer *15:* 99–108 (1975).
21 STRAUSSER, H. and HUMES, J.: Prostaglandin synthesis inhibition: effect on bone changes and sarcoma tumor induction in Balb/c mice. Int. J. Cancer *15:* 724–730 (1975).

Dr. O.J. PLESCIA, Waksman Institute of Microbiology, Rutgers, The State University of New Jersey, *New Brunswick, NJ 08903* (USA)

Morphological and Biological Features of MC-29 Virus-Induced Liver Tumors in Chicken

K. Lapis

1st Institute of Pathology, Semmelweis Medical University, Budapest

Liver cancers and transplantable hepatomas of different growth rates are the most widely used test objects in experimental cancer research [18]. Without exception, transplantable hepatomas used today have been established from primary liver cancers induced by various chemical carcinogens [19]. Though the etiology of human liver cancers is not yet clear, on the basis of numerous observations the pathogenic role of the viruses cannot be precluded. The connection between virus hepatitis and hepatoma is highly suggested by earlier and newer pathological findings. First of all, in tropical areas, where the rather frequent B type virus hepatitis is mainly responsible for the development of post-necrotic cirrhosis [4, 20, 21], the large majority of primary liver cancers arises on the ground of such macronodular liver cirrhosis [4, 5]. In Mali and Senegal, the infection with hepatitis B virus had been proved in more than 95% of the patients suffering from hepatoma [4].

By the recently introduced sensitive immunological methods – with the exception of America and northern Europe – the presence of $HB_s Ag$ could be detected with high frequency in the sera of patients with hepatoma [4]. It must also be considered that other viruses may play a role in the development of liver cancers, too.

In the light of the above, it is evident that concerning the pathogenesis of human liver cancers studies on transplantable hepatomas established from chemically induced liver cancers and transplantable virus-derived hepatomas are equally important.

In my present paper, I shall report on a new transplantable hepatoma induced and maintained in chickens. This deserves special attention because it is the first transplantable hepatoma derived from a virus-induced primary liver tumor [3, 13]. The interest toward this tumor may only be further increased as the primary liver cancer, from which this tumor originates, was not induced by hepatitis virus, but by the infection with the MC-29 virus strain originally known to cause fowl leukosis, namely myelocytomatosis. This shows that the spontaneous or induced variability of the viruses cannot only be manifested in the presence or absence of oncogenic or virogenic character, but also in the variability of the organotopy.

The avian MC-29 tumor virus strain has been isolated by IVANOV et al. [7] in Sofia from a hen with spontaneous leukosis. Later on, it was proved by BEARD et al. [2, 6, 13] that this virus strain is capable of producing a particularly wide spectrum of tumors, and that after i.v. infection primary liver tumor developed in a great proportion (30–50%) of baby chicks. The hepatomas always developed in cirrhosis-devoid livers within 30–40 days and led to the death of the animals [3, 13].

The tumors appeared as multiplex, circumscribed, greyish-white hemorrhagic nodules, 1–10 mm in diameter, which protruded above the liver surface (fig. 1a, b) [3, 13].

Histologically, the tumors can be ranged into the following 4 main groups: trabecular carcinoma, adenocarcinoma, anaplastic carcinoma, and so-called hemorrhagic carcinoma (fig. 2).

In one and the same liver, tumors of differing histological types occurred in varying combinations.

The tumor cells are characterized by a highly eosinophilic cytoplasm, a great, roundish-ovoid nucleus having a clear nucleoplasm and prominent nucleolus.

In the differentiated trabecular carcinomas, the ultrastructure and organization of tumor cells were very similar to that of the liver (fig. 3). The tumor cells had a large nucleus poor in chromatin and of low electron density, in which a prominent, hypertrophic nucleolus was located. The amount of cytoplasmic organelles and their relative numbers depended on the degree of differentiation of the tumor [3, 11]. In some of the tumor cells, decreased amounts of glycogen granules can still be seen. Numerous C-type virus particles are visible in the intercellular spaces. In the differentiated tumors, bile canaliculi regularly can be observed. Virus particles were often present in great numbers in the bile canaliculi and the 'budding' phenomenon could also occasionally be observed (fig. 4a, b).

Parallel with the dedifferentiation of the tumor, the fine structural features characteristic of the hepatic cells also diminished. The number of bile canaliculi and desmosomes as well as the occurrence of sinusoid-like structures gradually decreased.

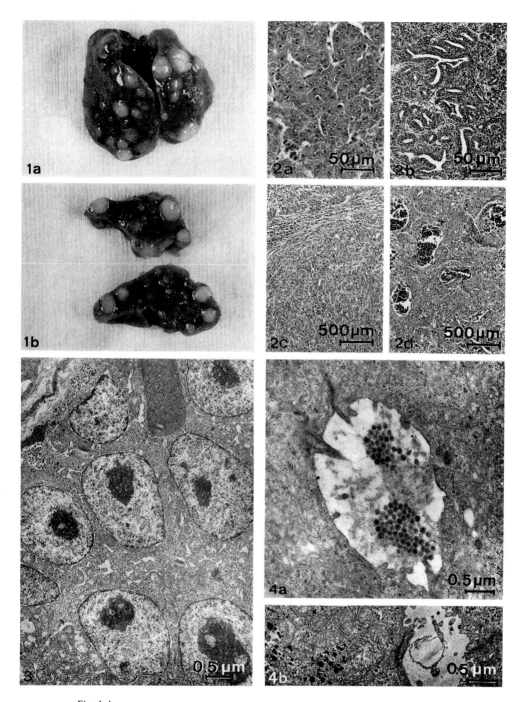

Fig. 1-4.

The transplantable hepatoma strain was derived in the following way from the virus-induced primary liver tumors discussed above [14, 15].

Pieces of tumorous tissue were i.p. inoculated with a trocar into 1-day-old chicks. Thereafter, the passages were performed s.c. and i.m. as well. During the first 3 passages, the animals were treated with 3×2.5 mg/kg cortisone. Later, the treatment with cortisone was no longer necessary, and the percentage of taking was 92%. We are at the 91st passage now and the tumor strain is being maintained in Hunnia hybrid chickens in our laboratory.

The tumors grew fast in the animals and within 10–12 days their size reached 15% of the body weight of the host (fig. 5).

The tumor-bearing animals died within 18 days in general. In about 25% of the animals, tumorous nodules were also detected in the liver, but it could not be decided with certainty whether they corresponded to metastases or represented primary liver tumors induced by viruses escaping from the transplant.

In the first few passages, the transplanted tumor exhibited glandular architecture histologically. Later on, the trabecular organization dominated and still later a tumor strain showing features of anaplastic liver cancer developed (fig. 6, 7).

During the first 10 passages, on the basis of the bile canaliculi, which were still easily discernible though they showed pathologic alterations and abortive features, it was still possible to ascertain the liver origin of the tumor by means of electron microscopic examination (fig. 8, 9) [11].

Subsequent passages resulted in a heavy dedifferentiation of hepatoma cells; bile canaliculi did not occur, not even in abortive form. The cells became poor in organelles. The amount of rough endoplasmic reticulum especially decreased, but the cells still had the enlarged nuclei and prominent nucleoli [12].

During the 2 years of maintenance, the cellular composition of the transplantable hepatoma became rather homogeneous and stabilized [12]. A few C-type virus particles occurred in the transplantable hepatoma strain even following maintenance for 2 years, and the phenomenon of 'budding' was also visible occasionally (fig. 9a, b).

Biochemical Studies

From the fifth passage, the biochemical character of the tumor was also studied [9]. The DNA content of the hepatoma proved to be twice as high

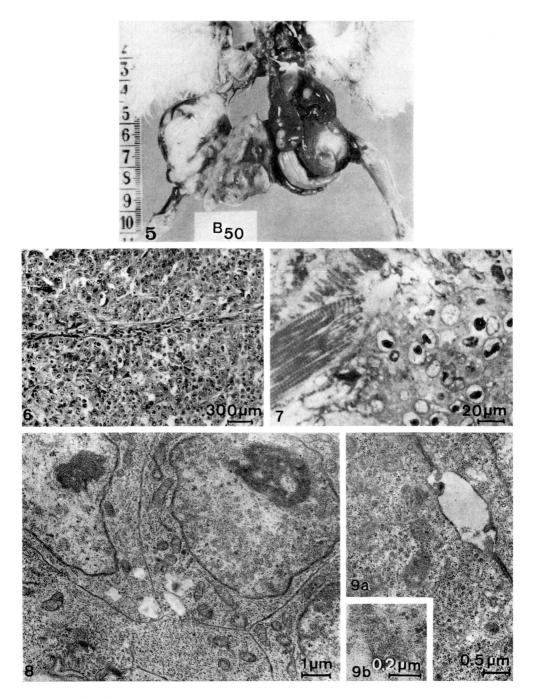

Fig. 5-9.

Table I. DNA, RNA, protein and phospholipid content of the whole tissue and the cellular fractions of the chicken liver and hepatoma

Fraction	DNA		RNA		Protein		Phospholipid	
	liver	hepatoma	liver	hepatoma	liver	hepatoma	liver	hepatoma
Whole tissue	100	*194*	100	*73*	100	*62*	100	*41*
Nucleus	100	*200*	100	*204*	100	*171*	–	–
15,000 g pellet	–	–	100	*34*	100	*46*	100	*20*
105,000 g pellet	–	–	100	*90*	100	*92*	100	*80*
105,000 g supernatant	–	–	100	*65*	100	*52*	100	*54*

Values are given as a percent of the normal chicken liver values.

Table II. Composition of chromatin in chicken liver and VTH hepatoma

Tissue	DNS	Histone	Nonhistone	0.35 M NaCl soluble fraction
Hepatoma	1.0	1.006	1.22	0.79
Normal liver	1.0	1.28	0.79	0.28

Results are related to 1 mg DNA of the corresponding tissue.

as that of the liver, while the RNA and protein content decreased. Further studies have shown that these alterations are the result of an increase in the nuclear DNA, RNA, protein contents, and of a decrease in the macromolecules in cytoplasmic organelles mainly in mitochondria and hyaloplasm (table I) [9].

Further on, it was examined whether the composition of the nuclear proteins differed from that of the liver or not. According to our findings, the ratio of the histone proteins to the DNA did not show significant changes in the hepatoma, while that of the non-histone protein increased in the tumor. An even higher increase could be detected in the tumor compared to norma liver in the case of the 0.35 M NaCl soluble protein fractions (table II) Studying the histone compounds of liver and hepatoma using polyacryl

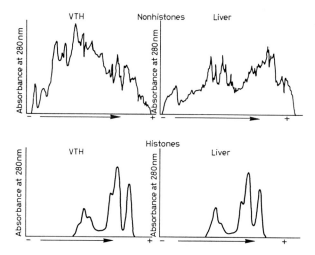

Fig. 10. Densitometric scans of SDS-polyacrylamide gel electrophoresis.

amide gel electrophoresis, no considerable difference was found between the two tissues; however, the nonhistone proteins differed in both quantity and quality (fig. 10). According to our findings, among the nonhistone proteins of the hepatoma, the quantity of the components of higher molecular weight (about 50,000 daltons) was dominating, while in the case of liver, approximately half of the protein molecules were situated in the faster migrating, lower molecular weight fractions [8]. Of course, further studies are needed. Nevertheless, it can be stated that in the case of this virus-derived hepatoma, both the ratio of nuclear proteins and the composition of the nonhistone proteins were altered in connection with the malignancy. As chromatin proteins have an important role in the regulation of gene expression, it is probable that various regulatory disorders may arise in consequence of the above alterations.

Therefore, it was examined whether the virus-derived transplantable hepatoma (VTH) had preserved some biochemical features which are known to be characteristic of the liver [9]. It has also been examined whether the high, elevated glucose-6-phosphate-dehydrogenase (G-6-PDH) activity characteristic of other transplantable hepatomas is also present in this fast-growing transplantable tumor. It was found that though in the tumor there were preserved G-6-Pase and aryl-hydrocarbon-hydroxilase activities characteristic of the differentiated function of the liver, the level of these enzymes

Table III. The activity of G-6-Pase, AHH and G-6-P DH in the chicken liver and hepatoma

	G-6-Pase nM P_i hour × mg protein	AHH pM 3-OH-BP hour × mg protein	G-6-P DH nM NADPH hour × mg protein
Liver % Hepatoma %a	1,400 ± 182 (100) 800 ± 96 (57)	30.0 ± 4.5 (100) 4.4 ± 2.3 (14.6)	21.3 ± 3.0 (100) 121.8 ± 8.6 (576)

G-6-Pase = glucose-6-phosphatase; AHH = aryl hydrocarbon hydroxilase;
G-6-P DH = glucose-6-phosphate dehydrogenase.
a Activities expressed as a percent of those found in the liver.

Table IV. Effect of methylcholantrene on the enzyme activities

	Treatment	AHH	G-6-Pase
Liver	–	100	100
Liver	MCa	528	90
Hepatoma	–	100	100
Hepatoma	MC	1	112

Results are expressed as a percent of the activities measured without stimulation.
a MC = 25 mg/kg methylcholantrene i.p. for 24 h.

was much decreased in comparison to that of the normal liver. In the tumor, the elevated G-6-PDH activity was also found (table III).

In the knowledge of the above-described quantitative changes in enzyme activity, we have examined whether the reactions of these enzymes to different inducers were the same in the tumor as in the liver. As enzyme inducers methylcholantrene, hydrocortisone, and insulin were applied.

The activities obtained after stimulation are expressed in the percentage of basic activity in the liver or hepatoma. 24 h after methylcholantrene administration, there was a more than 5-fold increase in the AHH activity of the liver, whereas in the hepatoma it decreased considerably (table IV).

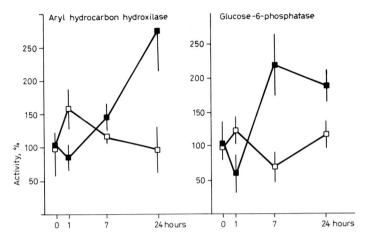

Fig. 11. Effect of hydrocortisone (250 mg/kg) on the enzyme activity of liver and hepatoma. ■ = liver; □ = hepatoma.

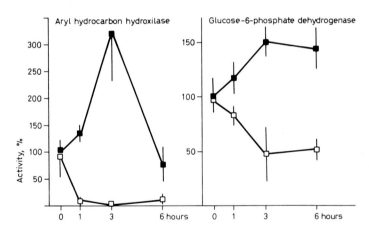

Fig. 12. Effect of insulin (20 IU/100 g) on the enzyme activity of liver and hepatoma. ■ = liver; □ = hepatoma.

In the case of hydrocortisone, following treatment, there was a considerable increase in AHH and G-6-Pase activity in the normal liver, while in the hepatoma the level of enzymes practically did not change (fig. 11).

3 h after treatment with insulin, both AHH and G-6-PDH activities showed a material increase in the liver, whereas in the hepatoma, the activities of these two enzymes fell below the level without induction (fig. 12).

Table V. Subcellular distribution of ^{3}H-hydrocortisone in the liver and hepatoma

	Liver dpm × 10^6 / grams	%		Hepatoma dpm × 10^6 / grams	%
Whole cell	4.536	100	$p<0.10$	3.762	100
Nucleus	1.942	42.80	$p<0.05$	1.274	33.80
Mitochondria	0.138	3.05	$p<0.10$	0.095	2.52
Microsome	0.034	0.75	$p<0.05$	0.038	1.02
Cytosol	2.421	53.40	$p<0.005$	2.354	62.60

Table VI. The binding of the cytoplasmic receptor complex of chicken liver and hepatoma to the liver and hepatoma DNA

	1 mg liver DNA		1 mg hepatoma DNA	
	liver cytosol	hepatoma cytosol	liver cytosol	hepatoma cytosol
Bound ^{3}H-Hydrocortisone pM/mg DNA	3.82	2.73	3.33	2.05

Summarizing the above observations, we found that on the effect of stimulators, severe regulatory disorders were manifested in the hepatoma. Two stimulators influenced enzyme activities in the hepatoma inversely to the normal liver and hydrocortisone did not show any effect at all. Therefore, in the following, we were trying to clarify why hydrocortisone is not capable of stimulating AHH and G-6-Pase in the hepatoma.

Examining the question whether the drug enters the cell [10], we have established that under *in vitro* circumstances, the hepatoma is capable of taking up nearly as much ^{3}H-hydrocortisone as the liver (table V). The incorporation study was carried out at 37°C and the highest ratio of isotope was found in cytosol and the nucleus.

Thus, it is evident that in the case of hepatoma, the labelled hydrocortisone enters both the cell and the nucleus.

In the following, we examined the binding of receptor-hydrocortisone complex to DNA according to the method recommended by BAXTER *et al.* [1]. We have stated [10] that the steroid-receptor complex of both liver and hepatoma binds better to the liver DNA than to the hepatoma DNA. The lowest rate of binding was obtained when attempting to associate hepatoma DNA to hepatoma cytosol-hydrocortisone complex (table VI). This fact could be one of the causes responsible for the lack of the effect observed after hydrocortisone treatment.

Summary

Further comparative studies on the biological and biochemical features of virus-derived transplantable and chemically induced hepatomas may contribute to the knowledge of human hepatomas. Evidence for the reprogramming of gene expression found in chemically induced transplantable hepatomas [22] was also found in this virus-derived hepatoma.

References

1 BAXTER, J.D.; ROUSSEAU, G.G.; BENSON, M.C.; GARCEA, R.L.; ITO, J., and TOMKINS, G.M.: Role of DNA and specific cytoplasmic receptors in glucocorticoid action. Proc. natn. Acad. Sci. USA *69:* 1892–1896 (1972).

2 BEARD, D.; CHABOT, J.F.; LANGLOIS, A.J.; HILLMAN, E.A., and BEARD, J.W.: Singularity of oncogenic activity of strain MC29 avian leukosis virus. Arch. Geschwulstforsch. *35:* 315–325 (1970).

3 BEARD, J.W.; HILLMAN, E.A.; BEARD, D.; LAPIS, K., and HEINE, U.: Neoplastic response of the avian liver to host infection with strain MC29 leukosis virus. Cancer Res. *35:* 1603–1627 (1975).

4 BLUMBERG, B.S.; LAROUZÉ, B.; THOMAS, W.; WERNER, B.; HESSER, J.E.; MILLMAN, I.; SAIMOT, G., and PAYET, M.: The relation of infection with the hepatitis B agent to primary hepatic carcinoma. Am. J. Path. *81:* 669–682 (1975).

5 DAVIES, J.N.P.: Hepatic neoplasm; in GALL and MOSTOFI The liver (by 34 authors). Int. Academy of Pathology, Monogr. Pathol., pp. 361–369 (Williams & Wilkins, Baltimore 1973).

6 HEINE, U.; MLADENOV, A.; BEARD, D., and BEARD, J.W.: Morphology of hepatoma induced by strain MC29 avian leukosis virus. Program 24th Annu. Meet. Electron Microscopy Society of America B-21 (1966).

7 IVANOV, X.; MLADENOV, Z.; NEDYALKOV, S.; TODOROV, T.G., and YAKIMOV, M.: Experimental investigations into avian leukoses. V. Transmission, haematology and morphology of avian myelocytomatosis. Bull. Inst. Pathol. Comp. Animaux Domest. *10:* 5–38 (1964).

8   JENEY, A. and KOVALSZKY, I.: Unpublished data.
9   KOVALSZKY, I.; ASBOTH, R.; JENEY, A., and LAPIS, K.: Biochemistry and enzyme induction in MC-29 virus-induced transplantable avian hepatoma. Cancer Res. *36:* 2140-2145 (1976).
10 KOVALSZKY, I.: Unpublished data.
11 LAPIS, K.: Fine structure of MC29 virus-induced liver tumours in chicken. III (in Hungarian only). Orvostudomány *25:* 277–301 (1974).
12 LAPIS, K.: IV. Fine structural features of the transplantable hepatoma derived from MC-29 virus-induced liver tumours in chicken (in Hungarian only). Orvostudomány *26:* 157–169 (1975).
13 LAPIS, K.; BEARD, D., and BEARD, J.W.: MC29 virus-induced liver tumours in chicken. I. Light microscopic morphology and histogenesis of virus-induced liver tumours (in Hungarian only). Orvostudomány *24:* 229–259 (1973).
14 LAPIS, K.; BEARD, D., and BEARD, J.W.: II. Transplantable hepatoma derived from MC-29 virus-induced liver tumour (in Hungarian only). Orvostudomány *25:* 151–160 (1974).
15 LAPIS, K.; BEARD, D., and BEARD, J.W.: Transplantation of hepatomas induced in the avian liver by MC29 leukosis virus. Cancer Res. *35:* 132–138 (1975).
16 LAPIS, K. and SCHAFF, Z.: Current stand of the virus aetiology of tumors (in Hungarian with Russian, German and English summary). Orv. Hetil. *115:* 1083–1091 (1974).
17 MELNYKOVYCH, G. and BISHOP, C.F.: Specific binding of cortisol in subcellular fractions of HeLa cells: temperature dependence and effects of inhibitors. Endocrinology *88:* 450–455 (1971).
18 MORRIS, H.P.: Studies on the development, biochemistry, and biology of experimental hepatomas; in HADDOW and WEINHOUSE Adv. Cancer Res., vol. 9, pp. 228–296 (Academic Press, New York 1965).
19 MORRIS, H.P. and WAGNER, B.P.: Induction and transplantation of rat hepatomas with different growth rate – including 'minimal deviation' hepatomas; in BUSCH Methods in cancer research, vol. 4, pp. 125–152 (Academic Press, New York 1968).
20 PAYET, M.; CAMAIN, R. et PENE, P.: Le cancer primitif du foie, étude critique à propos de 240 cas. Revue int. Hépat. *4:* 1–20 (1956).
21 STEINER, P.E. and DAVIES, J.N.P.: Cirrhosis and primary liver carcinoma in Uganda Africans. Br. J. Cancer *11:* 523–534 (1957).
22 WEBER, G.: Ordered and specific pattern of gene expression in differentiating and in neoplastic cells; in NAKAHARA, ONO, SUGIMURA and SUGANO Differentiation and control of malignancy of tumor cells. 4th Int. Symp. of the Princess Takamatsu Cancer Research Fund, pp. 151–180 (University of Tokyo Press, Tokyo 1974).

Dr. K. LAPIS, 1st Institute of Pathology, Semmelweis Medical University, *Budapest* (Hungary)

Immunoprevention of Leukemia in AKR Mice by Type-Specific Immune Gamma Globulin (IgG)

Robert J. Huebner, Paul J. Price, Raymond V. Gilden, Robert Toni, Richard W. Hill and Donald C. Fish

Laboratory of RNA Tumor Viruses, National Cancer Institute, Bethesda, Md.; Microbiological Associates, Torrey Pines Research Center, La Jolla, Calif., and Litton Bionetics, Inc., Frederick Cancer Research Center, Frederick, Md.

Introduction

In previous communications, we reported suppression of endogenous type C virogene expressions with the use of type-specific viral vaccines and immune gamma globulin (IgG) [1–4]. We now report significant prevention of spontaneous leukemia in AKR/J mice given 4 or 5 injections of antiviral IgG beginning at birth and continued to the 14th day of age in one experiment, and to the 20th day of age in a second experiment, the latter followed by additional immunization with AKR virus-specific vaccines [1]. Immunized and control AKR mice were then observed twice daily beginning at 170 days of age for development of spontaneous lymphocytic leukemias and/or thymic lymphomas. The earliest leukemias in AKR mice have been observed to begin at approximately 6 months of age.

Procedures

Since the procedures used were given in detail in a previous report [3], only a brief outline will be given here.

The anti-AKR antiviral IgG used for specific suppression of virogenes was prepared by injecting castrated male goats with 1,000× concentrated banded radiation leukemia virus (RadLV) of Lieberman and Kaplan [5] or banded Gross leukemia virus (GLV)[1] grown in SC-1 cells [6]. The goats,

[1] Heterotypic IgGs prepared against Rauscher leukemia virus (RLV), and AT124 immune xenotropic virus (MuX) were also produced in goats using the procedures as described. These viruses were unable to neutralize the ecotropic AKR virus either in the XC test or in the AKR tissues and were therefore regarded as a negative control.

12–18 months of age, were immunized intramuscularly at 1 to 2-week intervals with 1.0–3.0 mg of purified banded virus clarified by filtration (Millipore 1.2 μm) [7] mixed with equal parts of Freund's complete adjuvant (FCA). Animals were bled at 10-day intervals beginning 2 weeks after the third inoculation, following which the globulin fraction of pooled sera was precipitated by ammonium sulfate and resuspended in one half the original volume in PBS. After dialysis for 48 h at 4°C with several buffer changes, the IgG was filtered through a 0.45-μm filter, dispensed in vials and stored in nitrogen. Prior to use, it was heat-treated for 30 min at 56°C.

The anti-AKR neutralizing titers of the IgG pools were determined in the XC test using the SC-1 cell as the susceptible cell system [6]. IgG containing 1:1,600 anti-AKR virus-neutralizing units was given subcutaneously according to differing schedules in two experiments shown in table I. Virus titers in the hematopoietic tissues of the mice were assayed in the XC test with SC-1 cells employed as the susceptible cell system. At specified periods after the immunization period 2% wt/v extracts of tail segments (2 cm), which had been removed with a sharp scissors, were filtered, sonicated, and tested. Viral titers were given as the number of plaque-forming units per 0.4 ml expressed as a log titer.

Results

The XC titers present in 2% tail extracts from the immunized and control mice in two experiments are shown in table II. Table III shows the protection against spontaneous leukemia provided by the AKR virus-specific immunity. The experimental results revealed remarkable suppression of virus expression at 25 days and significant suppression up to 288 days of age. It was, therefore, apparent that normal levels of AKR virogene expressions were greatly reduced; this was particularly evident in the mice in experiment 1 wherein at 18–25 days of age the 32 immunized mice gave zero virus plaques while the controls had titers of $>10^{4.0}$. The log titer differences in the tails of the immunized compared to control mice in both experiments was in the order of 1,000–100,000 infectious units; also complement fixing tests of the tissues of mice in experiment 1 revealed less than 2 U of p30 antigen in spleens and thymuses of mice sacrificed at 25 days of age, while the same tissues in controls revealed final dilution titers of p30 of at least 1:800.

Some of the IgG-immunized mice in experiment 2 had virus titers at 25 days of $10^{2.6}$, thus reflecting lesser viral suppression by IgG given only 4 times

Table I. Dosage schedule of immunization

Experiment 1[1] (IgG only)		Experiment 2[2] (IgG + vaccines)	
Day 0[3]	0.05 ml	Day 0[3]	0.05 ml
Day 3	0.05 ml	Day 3	0.05 ml
Day 7	0.10 ml	Day 10	0.10 ml
Day 10	0.10 ml	Day 20	0.20 ml
		Day 25	0.20 ml SC[4]
			0.20 ml IP[4]
	GLV Vaccine		
		Day 39	0.40 ml IP[5]
		Day 53	0.40 ml IP[5]
	MSV (GLV) Vaccine	Day 63	0.10 ml SC[5]

[1] In experiment 1, 32 mice were immunized and 24 controls were not given any immunization. Addition controls not shown consisted of AKR mice given comparable IgG made against Rauscher leukemia virus (RLV) and murine xenotropic virus, neither of which had any detectable suppressive effects on AKR virus at 40 days of age.
[2] In experiment 2, 24 mice were immunized and 50 mice were held as controls. At 25 days of age, the mice given IgG were also given additional immunizations consisting of 3 injections of GLV vaccine at 14-day intervals, followed by MSV (GLV) live virus challenge 10 days later. The vaccine materials and immunization procedures were described in more detail in a previous report [1].
[3] Less than 24 h old.
[4] Administered with complete Freund's adjuvant.
[5] Administered without complete Freund's adjuvant.

Table II. Endogenous virus titers in 2% tail extracts from immunized and control mice

Immunized mice			Control mice		
age, days	number	log range	age, days	number	log range
Experiment 1					
18–25	32	$<10^{1.0}$ (all 32)	18–25	22	$10^{4.1}$ to $10^{5.1}$
238–248	24	$10^{3.5}$ to $10^{4.0}$	238–248	8	$10^{4.1}$ to $10^{5.5}$
Experiment 2					
25[1]	15	$<10^{1.0}$ to $10^{2.0}$	25	8	$10^{5.1}$ (all 8)
101–104	24	$10^{2.7}$ to $10^{4.1}$	101–104	23	$10^{3.6}$ to $10^{5.6}$
288	24[2]	$10^{1.3}$ to $10^{2.7}$	288	30	$10^{1.3}$ to $10^{4.1}$

[1] Results at age 25 are prior to vaccine administration.
[2] Mean log difference in titers = $10^{1.2}$.

Table III. Prevention of leukemia in AKR mice immunosuppressed by serotype-specific antiviral IgG antibodies. Mice at risk at 170+ days of age in test and control groups

	Age, days	Immunized (IgG) leukemia/totals (%)	Controls leukemia/totals (%)	Significance
Exp. 1	300	2/30 (6.6)	11/24 (45.8)	$p = 0.001$
	365	6/30 (20)	20/24 (83.3)	$p \ll 0.001$
Exp. 2[1]	250	0/24 (0)	9/50 (18)	$p = 0.02$
	300	1/24 (4.2)	30/50 (60)	$p \ll 0.001$

[1] The immunized mice in experiment 2 were also given killed GLV vaccine 5 days after completion of IgG treatment followed 10 days later by MSV (GLV) [1].

during a 20-day period. However, these mice at 25 days of age were given additional immunizations consisting of 3 inoculations of banded killed GLV vaccine at 14-day intervals [1] followed by MSV(GLV) pseudotype sarcoma virus administered 10 days after the last killed vaccine was given. The added immunization may have been responsible for lower than expected virus titers in the tails of the immunized mice when tested at 288 days, and it is not unlikely that the killed vaccine provided enhanced immunity. It is important to note that the type of immunity provided by both the IgG and the vaccines were of the same virus specificity.

Discussion and Conclusion

The ability to suppress endogenous type C viruses for relatively long periods with type C virogene specific IgG antisera[2] is in itself an extremely interesting observation. The possibility should be considered that similar procedures might prove applicable in suppressing other gene-related diseases.

The results of these two experiments establish a role for specific antiviral immunity in prevention of natural cancer in experimental animal systems; secondly, the data provide definitive evidence that the endogenous genetically inherited AKR virogenes [8] were causally responsible for the leukemias that characteristically occur in AKR mice; and, thirdly, these studies suggest additional experiments in which the antiviral specific IgG would be used in attempts to prevent sarcomas and carcinomas readily in-

[2] Antisera to heterotypic oncornaviruses did not suppress AKR virus.

duced in mice by carcinogenic chemicals. In support of the latter speculation, we can already report that subcutaneous sarcomas produced in C3H/f mice by 3-methylcholanthrene have been significantly reduced in number with the use of the same ecotropic virus specific IgG antibodies employed for prevention of leukemia.

Furth [9] and Law and Miller [10] reported protection against AKR lymphomas by removing the thymus, which appears to be the target organ for AKR lymphoma. Lunde and Gelderman [11] and Lemonde [12] reported variable but significant reductions in AKR leukemia following treatment with bacillus Calmette-Guerin (BCG) organisms. To our knowledge, this paper provides the first report of significant suppression of natural cancer with the use of type-specific anti-viral immunity to endogenous type C viruses.

Summary

AKR/J mice, 80–90% of which die of spontaneous lymphocytic leukemias by 12 months of age, were protected in two experiments from developing leukemia; in the first experiment, the mice were given a single course of treatment with anti-AKR viral antisera processed as immune gamma globulin (IgG) containing specific antibodies in high titer for type C AKR virus. The IgG injections which were given on the day of birth and on 4 subsequent days up to the 14th day of age, resulted in suppression of over 4 logs of normal AKR virogene expression up to 34 days of age; partial suppression persisted beyond 200 days of age. By 365 days, 6 of 30 (20%) of the immunized mice had developed fatal leukemias compared to 20 of 24 (83.3%) of the controls. In the second experiment, four IgG immunizations were given starting at birth and continued for 20 days; 5 days later, the mice were immunized 3 times 14 days apart with killed banded GLV vaccine which was then followed 10 days later by one injection of MSV(GLV). At 300 days of age, 30 of the 50 (60%) of controls were dead from leukemia whereas only 1 of 24 (4.2%) of the immunized mice had died of leukemia. Since a comparable heterotypic viral IgG (RLV) had no effect on virogene expressions in AKR mice, we concluded that these findings provide in classical fashion definitive evidence establishing the genetically transmitted AKR virogene expressions as the endogenous phenotypic cause of leukemia in the AKR mouse. The data also clearly show that the expressions of endogenous genetically transmitted virogenes and leukemia were both suppressed by serotype specific immunity.

References

1 HUEBNER, R.J.; GILDEN, R.V.; LANE, W.T.; TONI, R.; TRIMMER, R.W., and HILL, P.R.: Suppression of murine type-C RNA virogenes by type-specific oncornavirus vaccines: Prospects for prevention of cancer. Proc. natn. Acad. Sci. USA 73: 620–624 (1976).
2 HUEBNER, R.J.; GILDEN, R.V.; LANE, W.T.; TRIMMER, R.W., and HILL, P.R.: in CHIRIGOS Control of neoplasia by modulation of the immune system. pp. 381–389 (Raven Press, New York 1977).
3 HUEBNER, R.J.; GILDEN, R.V.; TONI, R.; HILL, P.R., and TRIMMER, R.W.: Proc. 3rd Int. Symp. on Detection and Prevention of Cancer (Dekker, New York, in press, 1976).
4 PRICE, P.J.; BELLEW, T.M.; KING, M.P.; FREEMAN, A.E.; GILDEN, R.V., and HUEBNER, R.J.: Prevention of viral-chemical co-carcinogenesis *in vitro* by type-specific anti-viral antibody. Proc. natn. Acad. Sci. USA 73: 152–155 (1976).
5 LIEBERMAN, M. and KAPLAN, H.S.: Leukemogenic activity of filtrates from radiation-induced lymphoid tumors of mice. Science 130: 387–388 (1959).
6 HARTLEY, J.W. and ROWE, W.P.: Clonal cell lines from a feral mouse embryo which lack host-range restrictions for murine leukemia viruses. Virology 65: 128–134 (1975).
7 OLPIN, J.; OROSZLAN, S., and GILDEN, R.V.: Biophysical-immunological assay for ribonucleic acid type C viruses. Appl. Microbiol. 28: 100–105 (1974).
8 CHATTOPADHYAY, S.K.; ROWE, W.P.; TEICH, N.M., and LOWY, D.R.: Definitive evidence that the murine C-type virus inducing locus *Akv-1* is viral genetic material. Proc. natn. Acad. Sci. USA 72: 906–910 (1975).
9 FURTH, J.: Prolongation of life with prevention of leukemia by thymectomy in mice. J. Geront. 7: 46–54 (1946).
10 LAW, L.W. and MILLER, J.H.: Observations on the effect of thymectomy on spontaneous leukemias in mice of the high-leukemic strains, RIL and C58. J. natn. Cancer Inst. 11: 253–262 (1950).
11 LUNDE, M.N. and GELDERMAN, A.H.: Resistance of AKR mice to lymphoid leukemia associated with a chronic protozoan infection, *Besnoitia jellisoni*. J. natn. Cancer Inst. 47: 485–488 (1971).
12 LEMONDE, P.: Protective effects of BCG and other bacteria against neoplasia in mice and hamsters. Natn. Cancer Inst. Monogr. 39: 21–29 (1973).

Dr. R.J. HUEBNER, Laboratory of RNA Tumor Viruses, National Cancer Institute, *Bethesda, MD 20014* (USA)

The Use of Mutagenicity Tests in Screening Chemical Carcinogens

R. MONTESANO

International Agency for Research on Cancer, Unit of Chemical Carcinogenesis, Lyon

The suggestion that cancer was due to a somatic mutation was put forward many years ago and this hypothesis received further support by the development of mutagenicity tests in which many carcinogens have been shown to be mutagenic. The inclusion of a metabolic activation system in these tests leads to this correlation, since it is now known that many carcinogens are not active as such but have to be metabolized into electrophilic metabolites which react with nucleophylic cellular components, such as DNA [MILLER, 1970]. These metabolites bind also to other cellular components such as RNA and proteins, and which of these molecules is the critical target in carcinogenesis is not known. With this in mind, these tests appear to be of use in detecting reactive molecules that may have a carcinogenic potential.

In this presentation, the N-nitroso compounds are considered in relation to three topics: correlation between mutagenic and carcinogenic effects; possibility to define potency in these two biological processes, and significance of these *in vitro* tests to the carcinogenic process in the entire organism. N-Nitroso compounds were chosen since knowledge of their metabolism, carcinogenic effects in various species, and mutagenicity in various test systems is rather well documented [MAGEE et al., 1975; MONTESANO and BARTSCH, 1976].

Correlation between Mutagenic and Carcinogenic Effects

In a review article by myself and H. BARTSCH [MONTESANO and BARTSCH, 1976], the mutagenicity of 54 N-nitroso compounds, including nitrosamines

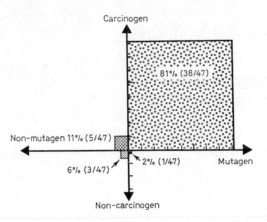

Fig. 1. Correlation between mutagenic and carcinogenic effects of N-nitroso compounds. From MONTESANO and BARTSCH [1976].

and nitrosamides, has been compared to their carcinogenic activity. Among the mutagenicity tests, we have included various test systems, namely the direct test, the *in vitro* tissue-mediated assay, the host-mediated assay, the dominant lethal test, induction of chromosome aberrations and tests in *Drosophila melanogaster*. Various genetic indicators were used in these systems. A major difference among these assays is that the metabolic activation of the compound examined is taken into account in some procedures, as in the tissue-mediated assay, but not in others, such as the direct test. The nitrosamines, as exemplified by DMN, are chemically stable under physiological conditions and it is now accepted that they exert their adverse biological effects after their metabolic conversion by microsomal mixed function oxidases to reactive intermediates. The nitrosamides, such as NMU, are unstable at physiological pH and decompose non-enzymatically to reactive, and in most cases, alkylating derivatives. As expected, the 24 *nitrosamines* examined were completely inactive when treated directly in a variety of cellular systems including phages, bacteria and yeast, apparently because these assays lack a metabolic activation system, and 24 out of 27 *nitrosamides* were active. On the other hand, 23 nitrosamines were tested in assay systems in which the genetic indicators were exposed to a mammalian metabolic activation system and 17 carcinogenic nitrosamines were mutagenic. The influence of modulators of drug-metabolizing enzymes, such as aminoacetonitrile, phenobarbitone and pregnenolone-16α-carbonitrile, on the mutagenicity of nitrosamines further stresses the importance of metabolism for the detection of this adverse biological effect.

In figure 1, the carcinogenic and mutagenic activities of the 47 N-nitroso compounds (23 nitrosamides and 24 nitrosamines), are plotted and 38 compounds were found to be carcinogenic and mutagenic. Five were not detected as mutagens, three non-carcinogens were non-mutagenic and only one compound, reported to be non-carcinogenic, exhibited a mutagenic effect [MONTESANO and BARTSCH, 1976].

A good correlation between carcinogenic and mutagenic effects was reported by MCCANN et al. [1975] and MCCANN and AMES [1976], who analyzed a much larger number of various classes of carcinogens and/or mutagens in a tissue-mediated mutagenicity assay utilizing S. typhimurium as a genetic indicator.

Mutagenic and Carcinogenic Potency

The various factors that may interfere in the determination of the mutagenic activity of a chemical are multiple and they are also influenced by other factors. These are mainly the balance between activation and detoxification processes, chemical stability and differential reactivity of the ultimate metabolite(s), bacterial metabolic activation, amount of microsomal enzymes, and mutagen specificity. Thus, it might become very difficult to express the results according to their degree of potency. Also, in the same mutagenicity test system as that employing the *Salmonella typhimurium* as genetic indicator in the presence of a metabolic activation system, the methodology used could play a major role. As shown by BARTSCH et al. [1976], the mutagenic effect of a series of nitrosamines with an aliphatic side chain of various lengths for S. typhimurium TA 1530 strain in the presence of rat liver microsomal fraction varied according to the type of incubation carried out. N-Nitroso derivatives of di-n-propylamine, di-n-butylamine and di-n-pentylamine showed a mutagenic effect in the plate incorporation assay but not in the liquid incubation system, whereas the dimethyl and diethylnitrosamine are effective in this latter system. This response was attributed to the different viability of the microsomal enzymes in these two experimental systems, as well as to the different reactivity of intermediate metabolites. Methodology also plays an important role in the detection of mutagenic activity of volatile compounds [BARTSCH et al., 1975; AMES et al., 1975].

In addition to these factors, the observation of the specificity of some chemicals to produce one type of mutation rather than another [KILBEY, 1976] makes it very hazardous to quantify the results in this test system.

A quantitative correlation between mutagenicity and carcinogenicity [MCCANN and AMES, 1976; TERANISHI et al., 1975] is an even more hazardous exercise, not only due to the problems inherent in the mutagenicity test itself, but also to the difficulty in determining the degree of potency in the carcinogenicity tests. The carcinogenic response to a chemical may vary greatly according to the species tested. For example, a dose level of 1 μg/kg of aflatoxin B_1 in the diet is sufficient to produce tumours of the liver in rats, whereas no tumours were detected in mice fed with a diet containing 1,000 μg/kg of aflatoxin B_1 [WOGAN, 1969; WOGAN et al., 1974]. A dose of more than 30 g per hamster is necessary to induce bladder cancer in approximately 50% of animals [SAFFIOTTI et al., 1967]. Both these compounds are carcinogenic in man. The polynuclear hydrocarbon, 7-H-dibenzo(c,g)carbazole, was found to be more carcinogenic for the hamster's respiratory tract [SELLAKUMAR and SHUBIK, 1972] than benzo(a)pyrene [MONTESANO et al., 1970], and an opposite response was observed when this compound was tested on the skin of mice.

Since the primary objective of carcinogenicity tests in animals is the primary prevention of human cancer, and since regulatory bodies are reluctant to accept these tests as evidence of a cancer risk for humans [MONTESANO and TOMATIS, 1977], major emphasis should be given to identifying chemicals in the environment with a carcinogenic effect. The classification of carcinogens by their potency in animals might lead to the introduction into the environment of chemicals which might subsequently prove to be highly carcinogenic in man.

It would be even more problematic to try to classify carcinogens according to their mutagenic effect, and only a better knowledge of the mechanism of action of chemical carcinogens might eventually allow a classification of the carcinogens by their potency.

Conclusion

Only some of these tests have been extensively investigated with a sufficient number of carcinogens and non-carcinogens, namely the tissue-mediated mutagenicity assays using *S. typhimurium* [MCCANN et al., 1975], and transformation *in vitro* of hamster embryo cells [PIENTA et al., 1977]. In these systems, a good correlation with the carcinogenicity data *in vivo* has been observed and they demonstrate the validity of these tests as preliminary screens for compounds that may be carcinogenic. However, these tests cannot substitute, at present, the long-term carcinogenicity assays in animals. The combination of these short-term tests with long-term carcinogenicity tests certainly represents a powerful tool for the control of environmental hazardous chemicals.

Regarding the prevention of human cancer, it is important that the public health authorities accept the results obtained in the long-term experimentation as evidence of a potential cancer risk for man. Otherwise, there is no reason to develop or improve short-term tests for carcinogens, the validity of which is based on their correlation with the *in vivo* data.

References

AMES, B.N.; MCCANN, J., and YAMASAKI, E.: Methods for detecting carcinogens and mutagens with the *Salmonella*/mammalian-microsome mutagenicity test. Mutation Res. *31:* 347–364 (1975).

BARTSCH, H.; CAMUS, A., and MALAVEILLE, C.: Comparative mutagenicity of N-nitrosamines in a semi-solid and in a liquid incubation system in the presence of rat or human tissue fractions. Mutation Res. *37:* 149–162 (1976).

BARTSCH, H.; MALAVEILLE, C., and MONTESANO, R.: Human, rat and mouse liver-mediated mutagenicity of vinyl chloride in *S. typhimurium* strains. Int. J. Cancer *15:* 429–437 (1975).

KILBEY, B.J.: Comparative aspects of mutagenesis in *Drosophila* and eukaryotic microorganisms. Hereditas *82:* 43–50 (1976).

MCCANN, J. and AMES, B.N.: Detection of carcinogens as mutagens in the *Salmonella* microsome test: assay of 300 chemicals. Proc. natn. Acad. Sci. USA *73:* 950–954 (1976).

MAGEE, P.N.; MONTESANO, R., and PREUSSMANN, R.: N-Nitroso compounds and related carcinogens; in SEARLE Chemical carcinogens, ACS Monogr. Ser. 173, vol. 11, pp. 491–625 (1976).

MILLER, J.A.: Carcinogenesis by chemicals: an overview – G.H.A. Clowes Memorial Lecture. Cancer Res. *30:* 559–576 (1970).

MONTESANO, R. and BARTSCH, H.: Mutagenic and carcinogenic N-nitroso compounds: possible environmental hazard. Mutation Res. *32:* 179–228 (1976).

MONTESANO, R.; BARTSCH, H., and TOMATIS, L. (eds): Screening tests in chemical carcinogenesis. IARC Scientific Publications No. 12 (International Agency for Research on Cancer, Lyon 1976).

MONTESANO, R.; SAFFIOTTI, U., and SHUBIK, P.: The role of topical and systemic factors in experimental respiratory carcinogenesis; in HANNA *et al.* Inhalation carcinogenesis, US AEC Symp. Ser. No. 18, pp. 353–371 (1970).

MONTESANO, R. and TOMATIS, L.: Legislation concerning chemical carcinogens in several industrialized countries. Cancer Res. *37:* 310–316 (1977).

PIENTA, R.J.; POILEY, J.A., and LEBHERZ, W.B., III: Morphological transformation of early passage golden Syrian hamster embryo cells derived from cryopreserved primary cultures as a reliable *in vitro* biossay for identifying diverse carcinogens. Int. J. Cancer. *19:* 642–655 (1977).

SAFFIOTTI, U.; CEFIS, F.; MONTESANO, R., and SELLAKUMAR, A.R.: Induction of bladder cancer in hamsters fed aromatic amines; in DEICHMAN and LAMPE Bladder cancer, pp. 129–135 (Aesculapius Publishing, Birmingham 1967).

SELLAKUMAR, A. and SHUBIK, P.: Carcinogenicity of 7H-dibenzo(c,g)carbazole in the respiratory tract of hamsters. J. natn. Cancer Inst. *48:* 1641–1646 (1972).

TERANISHI, K.; HAMADA, K., and WATANABE, H.: Quantitative relationship between the carcinogenicity and mutagenicity of polyaromatic hydrocarbons in S. *typhimurium* mutants. Mutation Res. *31:* 97–102 (1975).

WOGAN, G.N.: Naturally occurring carcinogens in foods. Prog. exp. Tumour Res., vol. 11, pp. 134–162 (Karger, Basel 1969).

WOGAN, G.N.; PAGLIALUNGA, S., and NEWBERNE, P.M.: Carcinogenic effects of low dietary levels of aflatoxin B_1 in rats. Food Cosmet. Toxicol. *12:* 681–685 (1974).

Dr. R. MONTESANO, International Agency for Research on Cancer, Unit of Chemical Carcinogenesis, *Lyon* (France)

Requirements for Tumor Antigen Immunogenicity

G. FORNI and G. CAVALLO

Istituto di Microbiologia, University of Torino, Torino

A specific cellular immune reactivity against autochthonous, syngeneic or allogenic tumors can be demonstrated *in vitro* in several host-tumor systems.

An overwhelming body of evidence indicates that thymus-dependent (T) lymphocytes play a major role in these cytotoxic activities against tumor-associated antigens (TATA). However, *in vivo* host cellular reactivity to TATA is often very weak, and its effectiveness appears to be impaired by multiple variables. Some of these are related to the age-dependent functional capability of the host's immune system [FORNI and COMOGLIO, 1973] and to the direct subversion of the immune system capability exerted by some tumors [PLESCIA *et al.*, 1975]. Another variable which can greatly modulate cellular reactivity to TATA is connected with the appearance of soluble serum factors that can potentiate or block host cellular reactivity in an evolving, complex way, especially in relation to tumor sizes [LANDOLFO *et al.*, 1977].

Various aspects of these 'environmental' variables affecting the post-recognitional phase of T cell reactivity to TATA are discussed in different parts of this book. Here, we will discuss some personal data and preliminary experiments of ours and deal particularly with the mechanisms involved in the specific proliferative recognition of TATA by syngeneic T lymphocytes.

Recent data have made it clear that at a post-recognition stage the presence of antigens of the major histocompatibility complex (MHC) is necessary for cytolysis. In effect, for an effective T cell lysis in mouse system, target cells must share part of the H-2 complex with the cells used in the sensitization phase. The findings of ZINKERNAGEL and DOHERTY [1974] and SHEARER *et al.* [1975] clearly show that thymus-derived cytotoxic lymphocytes generated both *in vivo* and *in vitro* to virus or hapten-modified autologous cells lyse only target cells which: (a) are modified by the same agent, and (b) express a common H2-K or H-2D alloantigen.

Identity at H-2K or H-2D alloantigens is also necessary for T cell lytic activity directed to different targets [BEVAN, 1975].

Furthermore, GERMAIN *et al.* [1975] and SCHRADER *et al.* [1975] have shown that by blocking the H-2K and H-2D alloantigens with antisera, the T cell lytic activity to TATA can be inhibited to a large extent. In a similar way, no T cell lytic activity is observed when target cells do not express H-2D or H-2K alloantigens [FORMAN and VITETTA, 1975]. The requirements for compatibility on the part of MHC between sensitizing and target cells for a specific T cell lysis have been determined mainly in the mouse system, but seem generalizable to guinea pig and human models.

While the post-activation requirements for a T cell-mediated cell lysis have been widely analyzed, the requirements for the recognitive T cell activation to different kinds of antigens are not so well defined. In mice and guinea pigs, it is well established that incongruity in some specialized products of the MHC (I region products) and other non-MHC-linked regions (M locus products) are required for cell proliferation in mixed lymphocyte culture (MLC). In both systems, antibodies masking the I region products abolish the recognitive T cell proliferation [MEO *et al.*, 1975; SCHWARTZ *et al.*, 1976].

In mouse systems, Ia-negative allogeneic cells [WAGNER *et al.*, 1975] or membrane-solubilized antigens [MANSON, 1976] are unable to induce a marked T cell proliferation by comparison with allogeneic lymphoid cells or lymphoid cell solubilized antigens. Moreover, while differences at H-2K and H-2D alloantigens seem to induce a moderate proliferation in MLR, SCHENDEL and BACH [1974] and WAGNER *et al.* [1975] have shown that differences at Ia alloantigens are necessary for the production of cytotoxic lymphocytes. In the guinea pig, only macrophages act as stimulator cells in MLC [GREINEDER and ROSENTHAL, 1975]. Moreover, recent studies have shown that antigen-pulsed macrophages exercise an obligatory role in the recognition of soluble antigens by immune T lymphocytes [ROSENTHAL and SHEVACH, 1973]. However, the activation of immune T lymphocytes by macrophage-associated soluble antigens requires both cells to share an identical Ia alloantigen. In this case, too, the antigen-presentation function of Ia compatible macrophage can be blocked by antisera containing antibodies to Ia-antigens [SHEVACH, 1976].

The relevance of these observations to an understanding of the requirements for a specific T cell recognition of TATA is evident. Since the recent evidence indicates that in several cases the operationally defined TATA are 'minor' alterations of normal membrane molecular structure, to a certain extent similar to membrane modifications induced by coupling with haptenic determinants, we built a rather simplified model [THOMAS et al., 1977]. Strain 2 (St. 2) inbred guinea pigs were sensitized to trinitrobenzene sulphonate (TNBS) by injection of TNBS emulsified in complete Freund's adjuvant. Two weeks later, when the animals presented a skin reactivity to TNBS, mineral oil-induced peritoneal exudate cells were obtained and T cells were purified by passing through a nylon wool column as described by SHEVACH et al. [1974]. The emerging populations enriched in T lymphocytes were incubated for 72 h *in vitro* with different mitomycin-C-treated stimulator cells. A strong TNP-specific proliferative recognition was obtained only by using TNP-coupled syngeneic or allogeneic leukocytes as stimulator. No TNP-specific proliferative response was observed when TNP groups were coupled to the membrane of different non-lymphoid, syngeneic or allogeneic cells, such as TNP-fibroblasts, TNP-sarcoma cells, and TNP-hepatoma cells.

Thus, all the cells unable to act as stimulator in MLC were unable to permit a TNP-specific proliferation by TNP-immune T lymphocytes. The T cell recognition of a modified membrane antigenic determinant take place only when the antigenic determinant is presented on a cell membrane which simultaneously carries specialized histocompatibility determinants stimulating in MLC. In general, if the TNP-induced cell membrane alteration can be compared to the relatively minor modifications of the membranes of the transformed cells functionally defined as TATA, it is possible to predict that a TATA will be effectually recognized by T lymphocytes only when it is expressed on a particular kind of cell type.

A more direct approach was made by analyzing some aspects of the relationships between Ia alloantigens and tumor transplantation antigen immunogenicity by comparing the antigenic and immunogenic properties of five mutant lines of St. 2 guinea pig leukemia (L2C) of B lymphocyte origin [FORNI et al., 1976]. These five lines were obtained from different laboratories in which L2C leukemia had been maintained and studied at various time. All these L2C lines bear two X-chromosomes and an identical extrametacentric chromosome, indicating a common ancestral origin. Moreover, they all bear surface IgM with an identical idiotypic determinant. All five lines have on their cell surface the B.1 specificity of the B alloantigen, which is the guinea pig equivalent of the murine H-2K and H-2D alloantigens. Four of these leukemic lines express on their surface the Ia.2 and Ia.4 alloantigens normally present on St.2 B lymphocytes. By contrast, in one line, Ia.2 and Ia.4 specificities were undetectable. It was observed that the four Ia-positive L2C lines have a readily demonstrable tumor antigen: syngeneic animals immunized with Ia-positive L2C cells in complete Freund's adjuvant were completely protected against a lethal challenge performed 14 days after immunization. By contrast, the Ia-negative L2C line did not possess tumor transplantation antigen, as measured by the same type of immunization protection test. To determine whether the deletion of Ia specificities was associated with the loss of tumor transplantation antigen, and not to an increased resistance of Ia-negative L2C cells to immunological attack, a criss-cross immunization and challenge experiment was performed. In St.2 guinea pigs, pre-immunized to the Ia-negative L2C leukemia and challenged with Ia-positive L2C leukemias, no protection or prolonged survival time rate was observed compared to the control groups, whereas St.2 animals immunized with the Ia-positive L2C lines were completely protected against the challenge with both the Ia-positive and Ia-negative L2C cells. These criss-cross protection experiments demonstrate that it is not simply that the Ia-negative leukemia lacks tumor transplantation antigen, but rather that the tumor transplantation antigen in this particular mutant cell line is not immunogenic; it can however be recognized by animals pre-immunized with Ia-positive L2C cells and bearing both tumor transplantation antigen and Ia alloantigens. These *in vivo* results along with other results obtained *in vitro* indicate that the Ia-negative L2C line shares a common tumor transplantation antigen with other L2C leukemias, and that it can be recognized by immunized syngeneic recipients. However, this antigen is not immunogeneic when it is present on Ia-negative L2C cells. A similar T lymphocyte activation pattern by L2C tumor antigen was observed *in vitro*. Here lymphocytes from untreated syngeneic donors did not develop a detectable response against L2C cells. However, anamnestic lymphocyte proliferative response to L2C cells was initiated in lymphocytes from St.2 animals which had been preimmunized with the four Ia-positive L2C lines. By contrast, no T cell anamnestic proliferative response to L2C was initiated in lymphocytes from St.2 animals preimmunized with the Ia-negative L2C cells.

These *in vivo* and *in vitro* experiments point to a relationship between the immunogenicity of L2C cells and the expression of Ia alloantigens. In this system, the immunization with the Ia-negative mutant line failed to elicit an effective recognition of tumor transplantation antigens, both *in vivo* and *in vitro*, suggesting that the presence of Ia alloantigens is a necessary requirement for the recognition of these TATA.

In another series of experiments, we evaluated some of the requirements for the *in vivo* recognition of allogeneic tumor cells in a mouse system. We used a spontaneous adenocarcinoma (ADK-lt) of Balb/c (H-2^d; Mls^b) origin which had been studied for long time in our laboratory. This tumor presents moderate antigenicity in syngeneic animals [CAVALLO and FORNI, 1974]. Earlier work had also indicated that the take of this tumor is influenced by spontaneous or artificially induced changes in the syngeneic host immune reactivity, suggesting that ADK-lt growth is hindered by a self-induced mechanism of immunological type [FORNI and COMOGLIO, 1973]. The ADK-lt cells present H-2K and H-2D alloantigens but do not bear the I region and M locus products. When the ADK-lt cells are obtained from tumor-bearing Balb/c mice, they are contaminated by a great number of infiltrating lymphoid cells [LANDOLFO, 1973], as usually happens with different solid tumors.

We have been able to establish a cloned line of this tumor *in vitro*, free of contaminating lymphoid cells. When groups of Balb/c mice were injected with equal amounts of ADK-lt cells obtained from bearing mice or from *in vitro* cultured cells, no differences in tumor growth and in survival time were observed.

In a similar way, when the experiment was performed in C57BL/6 mice (H-2_q; Mls^b), both kinds of cells were rapidly rejected without differences in the rejection pattern. On the contrary, when ADK-lt cells were injected in DBA/2 mice (H-2^d; Mls^a), a striking difference in the growth pattern was observed between cells obtained from tumor-bearing Balb/c mice and those cultured *in vitro*. The ADK-lt cells from tumor-bearing mice were rejected after a limited initial growth and after 14 days no DBA/2 mice still bore tumors. By contrast, injection of an equal number of ADK-lt cells cultured *in vitro* was followed by progressive tumor growth and death of the animals. However, when DBA/2 mice were challenged with the same number of ADK-lt cells cultured *in vitro* and mixed 3:1 with normal Balb/c spleen cells just before inoculation, the tumor cells were rapidly rejected with a growth rejection pattern identical to the one observed on injecting ADK-lt cells from tumor-bearing animals.

These preliminary observations seem to indicate that the presence of a target antigen of non-lymphoid tumor cells (I region and M locus products negative) is not enough to induce an effective cellular recognition. The simultaneous presence of I and M locus products positive spleen leukocytes will provide the necessary second signal for the recognition of ADK-lt target antigens.

The findings in the different experimental systems here discussed indicate that for a T cell recognitive activation, a TATA should be exposed on a cellular membrane carrying specific presentation structures. These structures appear to be restricted to histocompatibility alloantigens able to activate proliferation in MLC reactions. In guinea pigs, they appear to be mainly the I region products; in mice the I region and the M locus products. Thus, the actual immunogenicity of a membrane antigen acting as TATA appears to be much dependent on the presence on the membrane of histocompatibility alloantigens, as far as T cells are concerned.

An important limitation on the induction of a marked T cell reactivity to TATA to non-lymphoid tumors is that the I region and M locus products (or their equivalent in different species) are restricted to some particular kind of cells. However, our experiments suggest that collaboration between TATA and M locus products also take place if these antigens are presented simultaneously, but on different cells, consistently with the findings obtained in the 'three-cell' experiments [SCHENDEL and BACH, 1974; EIJSVOOGEL et al., 1973; SOLLINGER and BACH, 1976; ZARLING et al., 1976].

The biphasic involvement of different histocompatibility alloantigens in T cell reactivity seems to be a sophisticated recognitive system of cell membrane alteration. The presence of some alloantigens (I region and M locus products or their equivalent) is necessary for antigen recognition, while the presence of H-2K or H-2D (or their equivalents) is necessary for the antigen specific post-recognition T cell lytic activity.

As matters now stand, a wider scale verification of this model is a crucial point with a view to its possible application to the experimental or immunotherapeutical elicitation of a marked cell reactivity to tumors.

Acknowledgments

The experiments on guinea pig system here discussed were performed at the Laboratory of Immunology of the National Institute of Allergy and Infectious Diseases, Bethesda, Md. The experiments on mouse system were supported by a contract with the Italian National Council (CNR).

References

BEVAN, M.J.: The major histocompatibility complex determines susceptibility to cytotoxic T cells directed against minor histocompatibility antigens. J. exp. Med. *112:* 1347 (1975).

CAVALLO, G. and FORNI, G.: Cell reactivity towards syngeneic neoplastic cells in mice hypersensitized to dinitrophenol. Eur. J. Cancer *10:* 103 (1974).

EIJSVOOGEL, V.; BOIS, M. DU; MEINESZ, A.; BIERHOST, E.; ZEYLEMAKER, W., and SHEL-LESKENS, P.: The specificity and activation mechanism of cell mediated lympholysis (CML) in man. Transplantn Proc. *5:* 1675 (1973).

FORMAN, J. and VITETTA, E.S.: Absence of H-2 antigens capable of reacting with cytotoxic T cells on a teratoma line expressing T/t locus antigen. Proc. natn. Acad. Sci. USA *72:* 3661 (1975).

FORNI, G. and COMOGLIO, P.M.: Growth of syngeneic tumors in unimmunized newborn and adult hosts. Br. J. Cancer *27:* 20 (1973).

FORNI, G.; SHEVACH, E.M., and GREEN, I.: Mutant lines of guinea pig L2C leukemia. I. Deletion of Ia alloantigens is associated with a loss in immunogenicity of tumor-associated transplantation antigens. J. exp. Med. *143:* 1067 (1976).

GERMAIN, R.N.; DORF, M.E., and BENACERRAF, B.: Inhibition of T lymphocyte-mediated tumor-specific lysis by alloantisera directed against the H-2 serological specificities of the tumor. J. exp. Med. *142:* 1023 (1975).

GREINEDER, D.K. and ROSENTHAL, A.S.: Macrophage activation of allogeneic lymphocytes proliferation in guinea pig mixed lymphocyte culture. J. Immun. *114:* 1541 (1975).

LANDOLFO, S.: Monolayer cultures of transplantable tumors for cytotoxicity tests *in vitro*. Giorn. Batt. Vir. Immunol. *66:* 190 (1973).

LANDOLFO, S.; GIOVARELLI, M., and FORNI, G.: *In vitro* arming and blocking activity of sera from BALB/c mice bearing a spontaneous transplantable adenocarcinoma. Eur. J. Cancer (in press, 1977).

MANSON, L.A.: Intracellular localization and immunogenic capacities of phenotypic products of mouse histocompatibility genes. Biomembranes (in press, 1976).

MEO, T.; DAVID, C.S.; RIJNBEEK, A.M.; NABHOLZ, M.; MIGGIANO, V.C., and SHREFFLER, D.C.: Inhibition of mouse MLR by anti-Ia sera. Transplantn Proc. *7:* suppl. 1, p. 127 (1975).

PLESCIA, O.J.; SMITH, A.H., and GRINWICH, K.: Subversion of immune system by tumor cells and role of prostaglandins. Proc. natn. Acad. Sci. USA *72:* 1849 (1975).

ROSENTHAL, A.S. and SHEVACH, E.M.: Function of macrophages in antigen recognition by guinea pig T lymphocytes. I. Requirement for histocompatible macrophages and lymphocytes. J. exp. Med. *138:* 1194 (1973).

SCHENDEL, D.J. and BACH, F.H.: Genetic control of cell-mediated lympholysis in mouse. J. exp. Med. *140:* 1534 (1974).

SCHRADER, J.W.; CUNNINGHAM, B.A., and EDELMAN, G.M.: Functional interactions of viral and histocompatibility antigens of tumor surfaces. Proc. natn. Acad. Sci. USA *72:* 5066 (1975).

SCHWARTZ, R.H.; FATHMAN, C.G., and SACHS, D.: Inhibition of stimulation in murine mixed lymphocyte cultures with an alloantiserum directed against a shared Ia determinant. J. Immun. *116:* 929 (1976).

SHEARER, G.M.; REHN, T.G., and GARBARINO, C.A.: Cell-mediated lympholysis of trinitrophenil modified autalogous lymphocytes. Effector specificity to modified histocompatibility autigens. J. exp. Med. *141:* 1348 (1975).

SHEVACH, E.M.: The function of macrophages in antigen recognition by guinea pig T lymphocytes. III. Genetic analysis of the antigens mediating macrophage-T lymphocyte interaction. J. Immun. *116:* 1482 (1976).

Shevach, E.M.; Paul, W.E., and Green, I.: Alloantiserum induced inhibition of immune response gene product function. I. Cellular distribution of target antigens. J. exp. Med. *139:* 661 (1974).

Sollinger, H.W. and Bach, F.H.: Collaboration between *in vivo* response to LD and SD antigens of major histocompatibility complex. Nature, Lond. *259:* 487 (1976).

Thomas, D.W.; Forni, G.; Shevach, E.M., and Green, I.: The role of the macrophage as the stimulator cell in contact sensitivity. J. Immun. *118:* 1677 (1977).

Wagner, H.; Hammerling, G., and Rollinghoff, M.: Enhanced *in vitro* cytotoxic anti-H-2 responses in presence of Ia-antigens. Immunogenetics *2:* 257 (1975).

Zarling, J.M.; Raich, P.; McKeough, M., and Bach, F.H.: Generation of cytotoxic lymphocytes *in vitro* against autologous human leukemia cells. Nature, Lond. *262:* 691 (1976).

Zinkernagel, R.M. and Doherty, P.C.: Immunological surveillance against altered self components by sensitized T lymphocytes in lymphocytic choriomeningitis. Nature, Lond. *251:* 547 (1974).

Dr. G. Forni, Istituto di Microbiologia, University of Torino, Via Santena 9, *I-10126 Torino* (Italy)